OEN £17.00

Pitman Research Notes in Mathematics Series

AF868988

Submission of proposals for consideration
Suggestions for publication, in the form of outlines and representative samples, are invited by the Editorial Board for assessment. Intending authors should approach one of the main editors or another member of the Editorial Board, citing the relevant AMS subject classifications. Alternatively, outlines may be sent directly to the publisher's offices. Refereeing is by members of the board and other mathematical authorities in the topic concerned, throughout the world.

Preparation of accepted manuscripts
On acceptance of a proposal, the publisher will supply full instructions for the preparation of manuscripts in a form suitable for direct photo-lithographic reproduction. Specially printed grid sheets are provided and a contribution is offered by the publisher towards the cost of typing. Word processor output, subject to the publisher's approval, is also acceptable.

Illustrations should be prepared by the authors, ready for direct reproduction without further improvement. The use of hand-drawn symbols should be avoided wherever possible, in order to maintain maximum clarity of the text.

The publisher will be pleased to give any guidance necessary during the preparation of a typescript, and will be happy to answer any queries.

Important note
In order to avoid later retyping, intending authors are strongly urged not to begin final preparation of a typescript before receiving the publisher's guidelines and special paper. In this way it is hoped to preserve the uniform appearance of the series.

Longman Scientific & Technical
Longman House
Burnt Mill
Harlow, Essex, UK
(tel (0279) 26721)

Titles in this series

1 Improperly posed boundary value problems
A Carasso and A P Stone

2 Lie algebras generated by finite dimensional ideals
I N Stewart

3 Bifurcation problems in nonlinear elasticity
R W Dickey

4 Partial differential equations in the complex domain
D L Colton

5 Quasilinear hyperbolic systems and waves
A Jeffrey

6 Solution of boundary value problems by the method of integral operators
D L Colton

7 Taylor expansions and catastrophes
T Poston and I N Stewart

8 Function theoretic methods in differential equations
R P Gilbert and R J Weinacht

9 Differential topology with a view to applications
D R J Chillingworth

10 Characteristic classes of foliations
H V Pittie

11 Stochastic integration and generalized martingales
A U Kussmaul

12 Zeta-functions: An introduction to algebraic geometry
A D Thomas

13 Explicit *a priori* inequalities with applications to boundary value problems
V G Sigillito

14 Nonlinear diffusion
W E Fitzgibbon III and H F Walker

15 Unsolved problems concerning lattice points
J Hammer

16 Edge-colourings of graphs
S Fiorini and R J Wilson

17 Nonlinear analysis and mechanics: Heriot-Watt Symposium Volume I
R J Knops

18 Actions of fine abelian groups
C Kosniowski

19 Closed graph theorems and webbed spaces
M De Wilde

20 Singular perturbation techniques applied to integro-differential equations
H Grabmüller

21 Retarded functional differential equations: A global point of view
S E A Mohammed

22 Multiparameter spectral theory in Hilbert space
B D Sleeman

24 Mathematical modelling techniques
R Aris

25 Singular points of smooth mappings
C G Gibson

26 Nonlinear evolution equations solvable by the spectral transform
F Calogero

27 Nonlinear analysis and mechanics: Heriot-Watt Symposium Volume II
R J Knops

28 Constructive functional analysis
D S Bridges

29 Elongational flows: Aspects of the behaviour of model elasticoviscous fluids
C J S Petrie

30 Nonlinear analysis and mechanics: Heriot-Watt Symposium Volume III
R J Knops

31 Fractional calculus and integral transforms of generalized functions
A C McBride

32 Complex manifold techniques in theoretical physics
D E Lerner and P D Sommers

33 Hilbert's third problem: scissors congruence
C-H Sah

34 Graph theory and combinatorics
R J Wilson

35 The Tricomi equation with applications to the theory of plane transonic flow
A R Manwell

36 Abstract differential equations
S D Zaidman

37 Advances in twistor theory
L P Hughston and R S Ward

38 Operator theory and functional analysis
I Erdelyi

39 Nonlinear analysis and mechanics: Heriot-Watt Symposium Volume IV
R J Knops

40 Singular systems of differential equations
S L Campbell

41 N-dimensional crystallography
R L E Schwarzenberger

42 Nonlinear partial differential equations in physical problems
D Graffi

43 Shifts and periodicity for right invertible operators
D Przeworska-Rolewicz

44 Rings with chain conditions
A W Chatters and C R Hajarnavis

45 Moduli, deformations and classifications of compact complex manifolds
D Sundararaman

46 Nonlinear problems of analysis in geometry and mechanics
M Atteia, D Bancel and I Gumowski

47 Algorithmic methods in optimal control
W A Gruver and E Sachs

48 Abstract Cauchy problems and functional differential equations
F Kappel and W Schappacher

49 Sequence spaces
W H Ruckle

50 Recent contributions to nonlinear partial differential equations
H Berestycki and H Brezis

51 Subnormal operators
J B Conway

52 Wave propagation in viscoelastic media
F Mainardi
53 Nonlinear partial differential equations and their applications: Collège de France Seminar. Volume I
H Brezis and J L Lions
54 Geometry of Coxeter groups
H Hiller
55 Cusps of Gauss mappings
T Banchoff, T Gaffney and C McCrory
56 An approach to algebraic K-theory
A J Berrick
57 Convex analysis and optimization
J-P Aubin and R B Vintner
58 Convex analysis with applications in the differentiation of convex functions
J R Giles
59 Weak and variational methods for moving boundary problems
C M Elliott and J R Ockendon
60 Nonlinear partial differential equations and their applications: Collège de France Seminar. Volume II
H Brezis and J L Lions
61 Singular systems of differential equations II
S L Campbell
62 Rates of convergence in the central limit theorem
Peter Hall
63 Solution of differential equations by means of one-parameter groups
J M Hill
64 Hankel operators on Hilbert space
S C Power
65 Schrödinger-type operators with continuous spectra
M S P Eastham and H Kalf
66 Recent applications of generalized inverses
S L Campbell
67 Riesz and Fredholm theory in Banach algebra
B A Barnes, G J Murphy, M R F Smyth and T T West
68 Evolution equations and their applications
F Kappel and W Schappacher
69 Generalized solutions of Hamilton-Jacobi equations
P L Lions
70 Nonlinear partial differential equations and their applications: Collège de France Seminar. Volume III
H Brezis and J L Lions
71 Spectral theory and wave operators for the Schrödinger equation
A M Berthier
72 Approximation of Hilbert space operators I
D A Herrero
73 Vector valued Nevanlinna Theory
H J W Ziegler
74 Instability, nonexistence and weighted energy methods in fluid dynamics and related theories
B Straughan
75 Local bifurcation and symmetry
A Vanderbauwhede
76 Clifford analysis
F Brackx, R Delanghe and F Sommen
77 Nonlinear equivalence, reduction of PDEs to ODEs and fast convergent numerical methods
E E Rosinger
78 Free boundary problems, theory and applications. Volume I
A Fasano and M Primicerio
79 Free boundary problems, theory and applications. Volume II
A Fasano and M Primicerio
80 Symplectic geometry
A Crumeyrolle and J Grifone
81 An algorithmic analysis of a communication model with retransmission of flawed messages
D M Lucantoni
82 Geometric games and their applications
W H Ruckle
83 Additive groups of rings
S Feigelstock
84 Nonlinear partial differential equations and their applications: Collège de France Seminar. Volume IV
H Brezis and J L Lions
85 Multiplicative functionals on topological algebras
T Husain
86 Hamilton-Jacobi equations in Hilbert spaces
V Barbu and G Da Prato
87 Harmonic maps with symmetry, harmonic morphisms and deformations of metrics
P Baird
88 Similarity solutions of nonlinear partial differential equations
L Dresner
89 Contributions to nonlinear partial differential equations
C Bardos, A Damlamian, J I Díaz and J Hernández
90 Banach and Hilbert spaces of vector-valued functions
J Burbea and P Masani
91 Control and observation of neutral systems
D Salamon
92 Banach bundles, Banach modules and automorphisms of C*-algebras
M J Dupré and R M Gillette
93 Nonlinear partial differential equations and their applications: Collège de France Seminar. Volume V
H Brezis and J L Lions
94 Computer algebra in applied mathematics: an introduction to MACSYMA
R H Rand
95 Advances in nonlinear waves. Volume I
L Debnath
96 FC-groups
M J Tomkinson
97 Topics in relaxation and ellipsoidal methods
M Akgül
98 Analogue of the group algebra for topological semigroups
H Dzinotyiweyi
99 Stochastic functional differential equations
S E A Mohammed

100 Optimal control of variational inequalities
V Barbu
101 Partial differential equations and dynamical systems
W E Fitzgibbon III
102 Approximation of Hilbert space operators. Volume II
C Apostol, L A Fialkow, D A Herrero and D Voiculescu
103 Nondiscrete induction and iterative processes
V Ptak and F-A Potra
104 Analytic functions – growth aspects
O P Juneja and G P Kapoor
105 Theory of Tikhonov regularization for Fredholm equations of the first kind
C W Groetsch
106 Nonlinear partial differential equations and free boundaries. Volume I
J I Díaz
107 Tight and taut immersions of manifolds
T E Cecil and P J Ryan
108 A layering method for viscous, incompressible L_p flows occupying R^n
A Douglis and E B Fabes
109 Nonlinear partial differential equations and their applications: Collège de France Seminar. Volume VI
H Brezis and J L Lions
110 Finite generalized quadrangles
S E Payne and J A Thas
111 Advances in nonlinear waves. Volume II
L Debnath
112 Topics in several complex variables
E Ramírez de Arellano and D Sundararaman
113 Differential equations, flow invariance and applications
N H Pavel
114 Geometrical combinatorics
F C Holroyd and R J Wilson
115 Generators of strongly continuous semigroups
J A van Casteren
116 Growth of algebras and Gelfand–Kirillov dimension
G R Krause and T H Lenagan
117 Theory of bases and cones
P K Kamthan and M Gupta
118 Linear groups and permutations
A R Camina and E A Whelan
119 General Wiener–Hopf factorization methods
F-O Speck
120 Free boundary problems: applications and theory, Volume III
A Bossavit, A Damlamian and M Fremond
121 Free boundary problems: applications and theory, Volume IV
A Bossavit, A Damlamian and M Fremond
122 Nonlinear partial differential equations and their applications: Collège de France Seminar. Volume VII
H Brezis and J L Lions
123 Geometric methods in operator algebras
H Araki and E G Effros
124 Infinite dimensional analysis–stochastic processes
S Albeverio
125 Ennio de Giorgi Colloquium
P Krée
126 Almost-periodic functions in abstract spaces
S Zaidman
127 Nonlinear variational problems
A Marino, L Modica, S Spagnolo and M Degiovanni
128 Second-order systems of partial differential equations in the plane
L K Hua, W Lin and C-Q Wu
129 Asymptotics of high-order ordinary differential equations
R B Paris and A D Wood
130 Stochastic differential equations
R Wu
131 Differential geometry
L A Cordero
132 Nonlinear differential equations
J K Hale and P Martinez-Amores
133 Approximation theory and applications
S P Singh
134 Near-rings and their links with groups
J D P Meldrum
135 Estimating eigenvalues with *a posteriori/a priori* inequalities
J R Kuttler and V G Sigillito
136 Regular semigroups as extensions
F J Pastijn and M Petrich
137 Representations of rank one Lie groups
D H Collingwood
138 Fractional calculus
G F Roach and A C McBride
139 Hamilton's principle in continuum mechanics
A Bedford
140 Numerical analysis
D F Griffiths and G A Watson
141 Semigroups, theory and applications. Volume I
H Brezis, M G Crandall and F Kappel
142 Distribution theorems of L-functions
D Joyner
143 Recent developments in structured continua
D De Kee and P Kaloni
144 Functional analysis and two-point differential operators
J Locker
145 Numerical methods for partial differential equations
S I Hariharan and T H Moulden
146 Completely bounded maps and dilations
V I Paulsen
147 Harmonic analysis on the Heisenberg nilpotent Lie group
W Schempp
148 Contributions to modern calculus of variations
L Cesari
149 Nonlinear parabolic equations: qualitative properties of solutions
L Boccardo and A Tesei
150 From local times to global geometry, control and physics
K D Elworthy

151 A stochastic maximum principle for optimal control of diffusions
U G Haussmann
152 Semigroups, theory and applications. Volume II
H Brezis, M G Crandall and F Kappel
153 A general theory of integration in function spaces
P Muldowney
154 Oakland Conference on partial differential equations and applied mathematics
L R Bragg and J W Dettman
155 Contributions to nonlinear partial differential equations. Volume II
J I Díaz and P L Lions
156 Semigroups of linear operators: an introduction
A C McBride
157 Ordinary and partial differential equations
B D Sleeman and R J Jarvis
158 Hyperbolic equations
F Colombini and M K V Murthy
159 Linear topologies on a ring: an overview
J S Golan
160 Dynamical systems and bifurcation theory
M I Camacho, M J Pacifico and F Takens
161 Branched coverings and algebraic functions
M Namba
162 Perturbation bounds for matrix eigenvalues
R Bhatia
163 Defect minimization in operator equations: theory and applications
R Reemtsen
164 Multidimensional Brownian excursions and potential theory
K Burdzy
165 Viscosity solutions and optimal control
R J Elliott
166 Nonlinear partial differential equations and their applications. Collège de France Seminar. Volume VIII
H Brezis and J L Lions
167 Theory and applications of inverse problems
H Haario
168 Energy stability and convection
G P Galdi and B Straughan
169 Additive groups of rings. Volume II
S Feigelstock
170 Numerical analysis 1987
D F Griffiths and G A Watson
171 Surveys of some recent results in operator theory. Volume I
J B Conway and B B Morrel
172 Amenable Banach algebras
J-P Pier
173 Pseudo-orbits of contact forms
A Bahri
174 Poisson algebras and Poisson manifolds
K H Bhaskara and K Viswanath
175 Maximum principles and eigenvalue problems in partial differential equations
P W Schaefer
176 Mathematical analysis of nonlinear, dynamic processes
K U Grusa
177 Cordes' two-parameter spectral representation theory
D F McGhee and R H Picard
178 Equivariant K-theory for proper actions
N C Phillips
179 Elliptic operators, topology and asymptotic methods
J Roe
180 Nonlinear evolution equations
J K Engelbrecht, V E Fridman and E N Pelinovski
181 Nonlinear partial differential equations and their applications. Collège de France Seminar. Volume IX
H Brezis and J L Lions
182 Critical points at infinity in some variational problems
A Bahri
183 Recent developments in hyperbolic equations
L Cattabriga, F Colombini, M K V Murthy and S Spagnolo
184 Optimization and identification of systems governed by evolution equations on Banach space
N U Ahmed
185 Free boundary problems: theory and applications. Volume I
K H Hoffmann and J Sprekels
186 Free boundary problems: theory and applications. Volume II
K H Hoffmann and J Sprekels
187 An introduction to intersection homology theory
F Kirwan
188 Derivatives, nuclei and dimensions on the frame of torsion theories
J S Golan and H Simmons
189 Theory of reproducing kernels and its applications
S Saitoh
190 Volterra integrodifferential equations in Banach spaces and applications
G Da Prato and M Iannelli
191 Nest algebras
K R Davidson
192 Surveys of some recent results in operator theory. Volume II
J B Conway and B B Morrel
193 Nonlinear variational problems. Volume II
A Marino and M K Murthy
194 Stochastic processes with multidimensional parameter
M E Dozzi
195 Prestressed bodies
D Iesan
196 Hilbert space approach to some classical transforms
R H Picard
197 Stochastic calculus in application
J R Norris
198 Radical theory
B J Gardner
199 The C* – algebras of a class of solvable Lie groups
X Wang

200 Stochastic analysis, path integration and dynamics
D Elworthy
201 Riemannian geometry and holonomy groups
S Salamon
202 Strong asymptotics for extremal errors and polynomials associated with Erdös type weights
D S Lubinsky
203 Optimal control of diffusion processes
V S Borkar
204 Rings, modules and radicals
B J Gardner
205 Numerical studies for nonlinear Schrödinger equations
B M Herbst and J A C Weideman
206 Distributions and analytic functions
R D Carmichael and D Mitrović
207 Semicontinuity, relaxation and integral representation in the calculus of variations
G Buttazzo
208 Recent advances in nonlinear elliptic and parabolic problems
P Bénilan, M Chipot, L Evans and M Pierre
209 Model completions, ring representations and the topology of the Pierce sheaf
A Carson

Strong asymptotics for extremal errors and polynomials associated with Erdös-type weights

D S Lubinsky

Department of Mathematics,
Witwatersrand University,
P.O. Wits 2050,
Republic of South Africa,

and

Centre for Advanced Computing and Decision Support,
C.S.I.R.,
P.O. Box 395,
Pretoria 0001,
Republic of South Africa.

D S Lubinsky

Strong asymptotics for extremal errors and polynomials associated with Erdös-type weights

Longman Scientific & Technical

Copublished in the United States with
John Wiley & Sons, Inc., New York

Longman Scientific & Technical
Longman Group UK Limited
Longman House, Burnt Mill, Harlow
Essex CM20 2JE, England
and Associated Companies throughout the world.

Copublished in the United States with
John Wiley & Sons, Inc., 605 Third Avenue, New York, NY 10158

First published 1989

AMS Subject Classifications: (main) 42L05, 41A10, 41A17
(subsidiary) 41A25, 30E10

ISSN 0269-3674

British Library Cataloguing in Publication Data
Lubinsky, Doran
Strong asymptotic for extremal errors and polynomials associated with Erdos-type weights.
1. Extremal problems
I. Title
515′.64

ISBN 0-582-03779-4

Library of Congress Cataloging-in-Publication Data
Lubinsky, D.S. (Doron Shaul), 1955–
Strong asymptotics for extremal errors and polynomials associated with Erdos-type weights
p.cm.-- (Pitman research notes in mathematics series,
ISSN 0269-3674 ; 202)
Bibliography: p.
ISBN 0-470-21306-X
1. Extremal problems (Mathematics) 2. Orthogonal polynomials. 3. Asymptotic expansions
I. Title II. Title: Erdos-type weights. III. Series.
QA331.L8175 1988 515--dc19 88-24712 CIP

Printed and bound in Great Britain by
Biddles Ltd, Guildford and King's Lynn

Contents

Preface

In recent years, much attention has been devoted to the topic of L_p extremal polynomials associated with positive weights $W: \mathbb{R} \to \mathbb{R}$. The special case $p = 2$, which yields orthogonal polynomials with respect to the weight W^2, has attracted many researchers to this area.

In some recent lecture notes, E.B. Saff and the author considered **Freud weights**, that is weights of the form $W := e^{-Q}$, where $Q: \mathbb{R} \to \mathbb{R}$ is even and of smooth polynomial growth at infinity. Building on the work of others, we established Szegö type asymptotics for extremal errors of the form

$$E_{np}(W) := \min_{\deg(P)<n} \| \{x^n - P(x)\} W(x) \|_{L_p(\mathbb{R})},$$

as $n \to \infty$, together with the asymptotic behaviour in the plane of the corresponding extremal polynomials.

In this research note, we consider **Erdös weights**, that is, weights of the form $W := e^{-Q}$, where $Q: \mathbb{R} \to \mathbb{R}$ is even and of faster than polynomial growth at infinity. We obtain the behaviour as $n \to \infty$ of $E_{np}(W)$ for $1 \leq p \leq \infty$. For $1 < p \leq \infty$, we establish the asymptotic behavior in the plane of the associated extremal polynomials, and for $1 < p < \infty$, also their mean asymptotic behaviour on the real line.

In the special case $p = 2$, we establish pointwise asymptotics on the real line of orthonormal polynomials and Christoffel functions. The latter results depend heavily on certain one-sided, weighted polynomial approximations. We also provide inequalities for the largest zeros of extremal polynomials, and "sharp" Markov–Bernstein inequalities. In all cases, we provide rates, typically of $O(n^{-\delta})$, for some $\delta > 0$. As examples of weights to which the results are applicable, we mention

$W(x) := \exp(-\exp\{\log(A^2 + x^2)\}^{\alpha})$, $\alpha > 1$, A large enough, or, $W(x) := \exp(-\exp_k(|x|^{\alpha}))$, $\alpha > 0$, where, $k \geq 1$, and $\exp_k$ denotes the kth iterated exponential.

In an appendix, we also establish pointwise asymptotics on the real line of the orthonormal polynomials and Christoffel functions for Freud weights. The rate provided for the Christoffel functions reaches $O(n^{-1-\delta})$, for some $\delta > 0$, in some cases, and is uniform for real x in a certain range.

The author would like to acknowledge especially collaboration with, and the encouragement of, his mentors Eli Levin, Paul Nevai, Ed Saff and Colin Wright. Encouragement by, and earlier collaboration with Stan Bonan, Arnold Knopfmacher, Alphonse Magnus, Hrushikesh Mhaskar, and Vili Totik has helped this project, as have interesting insights and references from Jeff Geronimo, Attila Mate, Herbert Stahl, Joe Ullman, Walter Van Assche, and Matt Wyneken. Finally, many thanks to Moshe Sniedovich, Lyn Lochtenbergh, and Cobus Delport for helping to improve the author's T^3 skills in preparing this manuscript, and to Duncan Martin, for giving the author ample work time to complete it.

In memory of Israel Emmanuel Lubinsky

1 Notation

Throughout, C, C_1, C_2, ... denote positive constants independent of n and x, and also of any polynomial P. The same symbol does not necessarily denote the same constant in different occurrences. We use the usual o, O notation, and also use $\sim$ in the following sense:

$$f(x) \sim g(x),$$

if there exist C_1, $C_2 > 0$ such that

$$C_1 \leq f(x)/g(x) \leq C_2,$$

for the relevant range of x. Similar notations will be used for sequences and sequences of functions. The class of polynomials of degree at most n, with possibly complex coefficients, will be denoted by $\mathcal{P}_n$. The greatest integer $\leq$ x, will be denoted by <x>. Especially in Sections 9 to 15, δ, δ_1, $\hat{\delta}$, ϵ, η and so on will denote small positive numbers that often become smaller at each stage of the proof of a theorem. The symbol := or =: defines the quantity adjacent to the : .

Let $0 < p \leq \infty$ and let W(x) be a non–negative function on $\mathbb{R}$ such that

$$x^n W(x) \in L_p(\mathbb{R}),\ n = 0, 1, 2, \ldots . \tag{1.1}$$

For n = 1, 2, 3, ..., we define the <u>extremal error</u>

$$E_{np}(W) := \inf_{P \in \mathcal{P}_{n-1}} \| \{x^n - P(x)\}\ W(x) \|_{L_p(\mathbb{R})}, \tag{1.2}$$

and we define an nth L_p <u>extremal polynomial</u> $T_{np}(W,x)$ to be any monic polynomial of degree n satisfying

$$\| T_{np}(W,x) W(x) \|_{L_p(\mathbb{R})} = E_{np}(W). \tag{1.3}$$

Of course, $T_{np}(W,x)$ is unique for $1 \leq p \leq \infty$. We also make use of the **normalized extremal polynomial**

$$p_{np}(W,x) := T_{np}(W,x)/E_{np}(W), \tag{1.4}$$

which satisfies

$$\| p_{np}(W,x)W(x) \|_{L_p(\mathbb{R})} = 1. \tag{1.5}$$

In the case $p = 2$, $p_{np}(W,x)$ is an **orthonormal polynomial** for the weight $W^2(x)$. We shall also use the (standard) notation

$$p_n(W^2,x) := p_{n2}(W,x) = \gamma_n(W^2)x^n + \ldots, \tag{1.6}$$

where the **leading coefficient** is $\gamma_n(W^2) = 1/E_{n2}(W)$. Of course, we have for m, n = 0, 1, 2, ...,

$$\int_{-\infty}^{\infty} p_n(W^2,x)\, p_m(W^2,x)\, W^2(x)dx = \delta_{mn}. \tag{1.7}$$

The **nth Christoffel function** associated with W^2 is [23]

$$\lambda_n(W^2,x) := \inf_{P \in \mathcal{P}_{n-1}} \int_{-\infty}^{\infty} (PW)^2(t)\, dt \,/P^2(x),\ x \in \mathbb{R}, \tag{1.8}$$

$n = 1, 2, 3, \ldots$. When $p \geq 1$, it is known that $T_{np}(W,x)$ has n simple zeros, which we denote by

$$\infty > x_{1n}^{(p)} > x_{2n}^{(p)} > \ldots x_{nn}^{(p)} > -\infty.$$

The n+1 points of equioscillation of $T_{n\infty}(W,x)W(x)$ are denoted by

$$\infty > y_{1n} > y_{2n} > \ldots > y_{n+1,n} > -\infty.$$

If $xQ'(x)$ is positive and increasing in $(0, \infty)$, with limits 0 and ∞ at 0 and ∞ respectively, then for $u > 0$, we define **Freud's number** q_u to be the positive root of

$$u = q_u Q'(q_u), \tag{1.9}$$

and the **Mhaskar–Rahmanov–Saff** number a_u [19,20] to be the positive root of

$$u = \frac{2}{\pi} \int_0^1 a_u t\, Q'(a_u t)\, (1 - t^2)^{-1/2}\, dt. \tag{1.10}$$

Whenever defined, we set

$$A_u^* := \frac{2}{u\pi}\int_0^1 (a_u t)^2 Q''(a_u t)(1-t^2)^{-1/2}\,dt,\ u > 0, \tag{1.11}$$

and

$$\Delta_n := \left(\frac{\log n}{nA_n^*}\right)^{2/3},\ n = 1,2,3,\ \dots\ . \tag{1.12}$$

Given fixed $s \geq 0$, we shall often use the notation

$$c_n := a_n(1 + s\,\Delta_n),\ n = 1,2,3,\ \dots\ .$$

Further, whenever defined, we set

$$T(x) := (xQ'(x))'/Q'(x) = 1 + xQ''(x)/Q'(x), \tag{1.13}$$

$x \in (0,\infty)$. Let J be a fixed positive integer. We recall [11] the series

$$G_Q(x) := 1 + \sum_{n=J}^{\infty} (x/q_n)^{2n}\, n^{-1/2}\, e^{2Q(q_n)}, \tag{1.14}$$

and note that if we replace Q by Q/2, we obtain

$$G_{Q/2}(x) = 1 + \sum_{n=J}^{\infty} (x/q_{2n})^{2n}\, n^{-1/2}\, e^{Q(q_{2n})}. \tag{1.15}$$

The geometric mean of a non–negative function $w(x) \in L_1[-1,1]$ is

$$G[w] := \exp\left(\pi^{-1}\int_{-1}^{1} \log w(x)\,(1-x^2)^{-1/2}\,dx\right), \tag{1.16}$$

taken as 0 if the integral diverges to $-\infty$. The contracted geometric mean of a non–negative integrable function W(x) on $\mathbb{R}$ is

$$G[W(ax)] := \exp\left(\pi^{-1}\int_{-1}^{1} \log W(ax)\,(1-x^2)^{-1/2}\,dx\right), \tag{1.17}$$

$a > 0$. If $f(\theta)$ is a non–negative function on $[-\pi,\pi]$ for which $\log f \in L_1[-\pi,\pi]$, we define the Szegõ function [30,Ch.10]

$$D(f;z) := \exp\left(\frac{1}{4\pi}\int_{-\pi}^{\pi} \log f(t)\,\frac{1+ze^{-it}}{1-ze^{-it}}\,dt\right), \tag{1.18}$$

$|z| < 1$, satisfying [30,p.276]

$$|D(f;e^{i\phi})|^2 = f(\phi),\ \text{a.e. } \phi \in [-\pi,\pi], \tag{1.19}$$

the left–hand side being a radial or non–tangential limit. Given a function $F: [-1,1] \to [0, \infty)$ for which $\log F(\cos t) \in L_1[-\pi,\pi]$, we shall often use the notation

$$D(F(\cos t);z) := \exp\left(\frac{1}{4\pi}\int_{-\pi}^{\pi} \log F(\cos t)\, \frac{1 + ze^{-it}}{1 - ze^{-it}}\, dt\right), \tag{1.20}$$

$|z| < 1$, and for $\phi \in (-\pi,\pi)$, for which the integral is defined,

$$\Gamma(F;\phi) := \frac{1}{4\pi}\int_{-\pi}^{\pi} \{\log F(\cos t) - \log F(\cos \phi)\}\cot\left(\frac{\phi - t}{2}\right)dt. \tag{1.21}$$

Finally,

$$\varphi(z) := z + (z^2 - 1)^{1/2},\ z \in \mathbb{C}\backslash[-1,1], \tag{1.22}$$

denotes the conformal map of $\mathbb{C}\backslash[-1,1]$ onto $\{z: |z| > 1\}$, the branch of the square root being the usual one, and we set

$$\sigma_p := \begin{cases} \{\Gamma(1/2)\ \Gamma((p+1)/2)\ /\ \Gamma(p/2+1)\}^{1/p}, & 0 < p < \infty, \\ 1 & , p = \infty. \end{cases} \tag{1.23}$$

Given a non–negative finite measure $d\mu(\theta)$ on $[-\pi,\pi]$, whose support contains infinitely many points, for $j = 0, 1, 2, \ldots$, we denote the jth orthonormal polynomial for $d\mu$ by

$$\phi_j(d\mu,z) = \kappa_j(d\mu)\, z^j + \ldots \in P_j,\ \kappa_j(d\mu) > 0, \tag{1.24}$$

satisfying

$$\frac{1}{2\pi}\int_{-\pi}^{\pi} \overline{\phi_j(d\mu,z)}\, \phi_k(d\mu,z)\, d\mu(\theta) = \delta_{jk},\quad z = e^{i\theta}. \tag{1.25}$$

When $d\mu(\theta) = f(\theta)\, d\theta$, we write $\phi_j(f,z)$, $\kappa_j(f)$, and so on. At various stages of this paper, we shall introduce more notation, noticeably in Section 4.

2 Statement of Main Results

We shall not attempt to survey the subject of orthogonal or extremal polynomials for weights on $\mathbb{R}$, but shall rather refer the reader to [23], [10], [12], [14,Ch.1(i)]. The results of this paper are closely related to those in [14], where Freud weights were treated. Roughly speaking, a **<u>Freud weight</u>** has the form $W(x) := e^{-Q(x)}$, $x \in \mathbb{R}$, where $Q(x)$ is even and continuous in $\mathbb{R}$, and is of polynomial growth at infinity.

In this paper, we treat **<u>Erdös weights</u>**, which have the form $W(x) := e^{-Q(x)}$, where $Q(x)$ is even, continuous in $\mathbb{R}$, and grows faster than any polynomial at infinity. We call them Erdös weights, for Erdös [3] was the first to consider them, obtaining the contracted zero distribution of their orthogonal polynomials, as well as the asymptotic behaviour of the largest zero thereof. Erdös also considered the non–even case. Asymptotics of the coefficients in the three–term recurrence relation for the orthonormal polynomials for these weights were obtained in [7], and their Markov–Bernstein inequalities were investigated in [13]. Other references relevant to Erdös weights are [20,31].

The new features of this paper, in comparison to [14], are the development of certain one–sided weighted approximations. These enable us to prove pointwise asymptotics on the real line for Christoffel functions, and orthogonal polynomials, as well as to extend the range of p for which some of the asymptotics in [14] were established. The application of these ideas to Freud weights is outlined in the appendix.

In constructing the one–sided approximations, we draw heavily on ideas in [14], on sharper asymptotics for the entire function G_Q (introduced in [11]), and also establish some elementary technical lemmas.

In order to state our main results, we need the following class of weights:

Definition 2.1

(a) Let $W := e^{-Q}$, where Q is even and continuous in $\mathbb{R}$; $Q''(x)$ is continuous in $(0, \infty)$; $Q'(x)$ is positive in $(0, \infty)$, and $T(x)$ (defined by (1.13)) is positive and increasing in $(0, \infty)$, with

$$T(0+) := \lim_{x \to 0+} T(x) > 0, \tag{2.1}$$

while

$$\lim_{x \to \infty} T(x) = \infty. \tag{2.2}$$

Assume further that

$$T(x) = O(Q'(x)^{1/15}), \ x \to \infty, \tag{2.3}$$

and

$$Q''(x)/Q'(x) \sim Q'(x)/Q(x) \ , \ x \text{ large enough}. \tag{2.4}$$

Then we shall call W a **smooth Erdős weight**, and shall write $W \in SE$.

(b) Let $j \geq 3$. We write $W \in SE(j)$ if $W \in SE$, and $Q^{(j)}(x)$ exists for x large, while for $k = 3, 4, \ldots j$,

$$|Q^{(k)}(x)| / Q'(x) \leq C \, (Q'(x)/Q(x))^{k-1}, \ x \text{ large enough}. \tag{2.5}$$

Remarks

(a) Let

$$\exp_0(x) := x, \ x \in \mathbb{R},$$

$$\exp_k(x) := \exp(\exp_{k-1}(x)), \ x \in \mathbb{R}, \ k = 1, 2, 3, \ldots ;$$

and let

$$\log_0 x := x, \ x \in \mathbb{R},$$

$$\log_k x := \log(\log_{k-1} x),\ x > \exp_{k-1}(0),\ k = 1, 2, 3, \ldots .$$

Further, for $\alpha > 0$ and $k \geq 1$, let

$$Q_{k,\alpha}(x) := \exp_k(|x|^{\alpha}),\ x \in \mathbb{R}, \tag{2.6}$$

and let

$$W_{k,\alpha}(x) := \exp(-Q_{k,\alpha}(x)),\ x \in \mathbb{R}.$$

Then $W \in SE$ and $W \in SE(j)$ for all $j \geq 3$. In this case [13],

$$T(x) = \alpha \{\prod_{j=1}^{k} \log_j Q_{k,\alpha}(x)\} \{1 + o(1)\},\ x \to \infty. \tag{2.7}$$

Another example is

$$W(x) := \exp(-\exp[\{\log(A + x^2)\}^{\alpha}]),\ x \in \mathbb{R},$$

where $\alpha > 1$, and A is large enough. This W belongs to SE and SE(j) for all $j \geq 3$. These two examples illustrate the fact that, given any growth function Q of sufficient smoothness and regularity, and of faster than polynomial growth at infinity, SE and SE(j) will contain $W := e^{-Q}$.

(b) Probably all the results of this paper remain valid under the slightly weaker condition on T that

$$T(x) \leq C\, T(y),\ 0 < x < y.$$

Further, (2.3) can be weakened (in the sense that 1/15 can be increased) in many of the results. In any event, it is an extremely weak regularity condition, as (2.7) shows. One typically has

$$T(x) = O(\log Q'(x))^{1+\epsilon},\ x \to \infty, \text{ for each } \epsilon > 0.$$

(c) The class SE overlaps the class VSE considered in [7], but neither contains the other. SE allows slightly more irregular growth of T(x) for large x.

First we give a result on L_p extremal errors and polynomials:

Theorem 2.2

Let $W \in SE(3)$.

(a) Given $1 \le p \le \infty$, there exists $\delta > 0$, such that as $n \to \infty$,

$$E_{np}(W) = 2\sigma_p \, (a_n/2)^{n+1/p} \, G[W(a_n x)] \, \{1 + O(n^{-\delta})\}. \tag{2.8}$$

(b) Let $1 < p \le \infty$, and for n large enough, let

$$F_n(x) := W(a_n x) \, (1 - x^2)^{1/(2p)}, \; x \in [-1,1]. \tag{2.9}$$

Then there exists $\delta > 0$, such that uniformly for z in closed subsets of $\mathbb{C}\backslash[-1,1]$, as $n \to \infty$,

$$p_{np}(W, a_n z) \,/\, \{a_n^{-1/p} \, \varphi(z)^n \, D^{-2}(F_n(\cos t); \tfrac{1}{\varphi(z)})\}$$
$$= (2\sigma_p)^{-1} \, \{1 + O(n^{-\delta})\}. \tag{2.10}$$

(c) Let $1 < p < \infty$, and for $n \ge 1$, let

$$c_n := a_n(1 + s\Delta_n), \tag{2.11}$$

where Δ_n is defined by (1.12), and s is a fixed, but large enough, positive number. For $n \ge 1$, let

$$\hat{F}_n(x) := W(c_n x) \, (1 - x^2)^{1/(2p)}, \; x \in [-1,1]. \tag{2.12}$$

Then there exists $\delta > 0$ such that as $n \to \infty$,

$$\left\| \left\{ c_n^{1/p} p_{np}(W, c_n x) - \frac{1}{\sigma_p} \operatorname{Re}[\, z^n D^{-2}(\hat{F}_n(\cos t); \tfrac{1}{z}) \,] \right\} W(c_n x) \right\|_{L_p[-1,1]} = O(n^{-\delta}), \tag{2.13}$$

where for $x = \cos\theta \in [-1,1]$, we set $z := e^{i\theta}$.

Proof

(a) See Theorem 12.1.

(b) See Theorem 12.2 for $2 \le p \le \infty$, and Theorem 14.5 for $1 < p < 2$.

(c) See Theorem 14.3 and set $m = n$ there. □

Remarks

(a) Theorem 12.1 also contains an asymptotic upper bound for $E_{np}(W)$, $0 < p < 1$.

(b) Theorem 12.2 also lists an asymptotic formula for the non–normalized extremal polynomial $T_{np}(W, a_n z)$. Note that δ is independent of the particular closed subset.

(c) Since in all our weighted approximations (Theorems 9.1 and 10.1), there is a free polynomial of degree $O(n\Delta_n) \geq n^{1/4}$ at the nth stage, one can multiply by other factors, and still obtain results such as Theorem 2.2. This is especially true in the case $p = 2$, of orthogonal polynomials (cf. Theorems 3.4, 3.5 in [14,pp.20–22]). However, one loses the rate $O(n^{-\delta})$, and therefore we omit these results for simplicity.

(d) When $Q(x)$ grows so fast that

$$\lim_{n\to\infty} a_n/a_{n^{1/2}} = 1, \tag{2.14}$$

(for example $Q_{k,\alpha}$ of (2.6) satisfies this for $k \geq 2$, and $\alpha > 0$), and Q'' is continuous at 0, then we can prove the above result for $W \in SE$, so that the existence of Q''' is not required. One loses the rate of convergence, however. We very briefly outline the proof. The main ingredient of Theorem 2.2 is the one–sided approximations of Theorem 10.1. One needs Q''' there only in passing from $1/G_{Q/2}$ to W. However, when (2.14) holds, one can approximate $Q(a_n x)$ by polynomials of degree $n^{1/4+}$ for $|x| \leq a_{n^{1/4}}/a_n$ using just Jackson's theorem, and then compose these polynomials with the $n^{1/2}$–th partial sums of the Maclaurin series of e^x, to obtain suitable polynomial approximations for $W^{-1}(a_n x)$. The intervals $|x| \in [a_{n^{1/4}}/a_n, 1]$ can be filled using Lemma 4.6.

(e) Note that when $W \in SE(3)$, then

$$\Delta_n \sim \left(\frac{\log n}{nT(a_n)}\right)^{2/3}, \quad n \to \infty,$$

(see Lemma 3.4(c) below). Thus in Theorems 2.2 and 2.4, one may choose a more explicit form for c_n, namely,

$$c_n := a_n \left(1 + s\left(\frac{\log n}{nT(a_n)}\right)^{2/3}\right),$$

with s fixed, but large enough. In any event,

$$\Delta_n = o\left(\left(\frac{\log n}{n}\right)^{2/3}\right), \ n \to \infty,$$

so that

$$c_n/a_n = 1 + o\left(\left(\frac{\log n}{n}\right)^{2/3}\right), \ n \to \infty.$$

It seems non–trivial to replace c_n in (2.12) and (2.13) by a_n. This was performed for the Christoffel functions (see Theorem 15.3), but was lengthy and led to loss of precision.

(f) Theorem 2.2 is the analogue for Erdös weights of Theorems 3.1 and 3.2 in [14,pp.17–18]. The new features above as compared to [14], are the L_p mean asymptotics (2.13), and also the proof of (2.10) for $1 < p < 2$. In [14], (2.10) was established for Freud weights only for $2 \le p \le \infty$. The new methods used here may equally well be applied to fill these gaps for Freud weights, and draw on ideas used in similar problems for weights on [–1,1] [15].

(g) The term Re[z^n ...]$W(c_nx)$ in (2.13) may also be expressed as $\cos(n\theta + \Gamma(\hat{F}_n^2;\theta))$, where $\Gamma(\cdot\,;\cdot\,)$ is defined by (1.21) – compare Theorem 2.4 below.

Next, we give asymptotics for Christoffel functions:

Theorem 2.3

Let $W \in SE(3)$, and let Q' be continuous at 0, satisfying there a Lipschitz condition of some positive order. Let F_n be given by (2.9) with $p = 2$. Then there exists $\delta > 0$ such that as $n \to \infty$, uniformly for

$$|x| \le 1 - n^{-\delta}, \ x = \cos \phi, \ z := e^{i\phi}, \tag{2.15}$$

we have

$$\pi\,\lambda_n^{-1}(W^2, a_n x)\,F_n^2(x)\,a_n/n$$

$$= 1 + n^{-1}\,\mathrm{Re}\,[zD'(F_n^2(\cos t);z)/D(F_n^2(\cos t);z)] + o(\epsilon_n), \tag{2.16}$$

$$= 1 + n^{-1}\,\Gamma'(F_n^2;\phi) + o(\epsilon_n), \tag{2.17}$$

where

$$\epsilon_n := n^{-19/58},\ n \geq 1. \tag{2.18}$$

If also $W \in SE(4)$, then we may choose

$$\epsilon_n := n^{-42/87},\ n \geq 1. \tag{2.19}$$

For $n \geq 1$, and $x \in (-1,1)$, let

$$\mu_{n,a_n}(x)$$

$$:= 2\pi^{-2}\int_0^1 \frac{(1-x^2)^{1/2}}{(1-s^2)^{1/2}}\,\frac{(a_n s Q'(a_n s) - a_n x Q'(a_n x))}{n(s^2 - x^2)}\,ds. \tag{2.20}$$

Then

$$\mu_{n,a_n}(x) > 0,\ x \in (-1,1), \tag{2.21}$$

and

$$\int_{-1}^1 \mu_{n,a_n}(x)\,dx = 1. \tag{2.22}$$

Further, as $n \to \infty$, uniformly for the range (2.15), we may also write

$$\lambda_n^{-1}(W^2, a_n x)\,W^2(a_n x)\,a_n/n = \mu_{n,a_n}(x) + o(\epsilon_n n^{\delta/2}). \tag{2.23}$$

Here $\delta > 0$ may be made arbitrarily small.

<u>Proof</u>

See Theorem 15.5. □

<u>Remarks</u>

(a) If we replace (2.3) by the more realistic restriction

$$T(x) = O(\{\log Q'(x)\}^2),\ x \to \infty,$$

then the order estimates above can be substantially improved.

(b) Usually asymptotics such as (2.16) are given for the more restrictive range $|x| \le 1 - \delta$, $\delta \in (0,1)$ fixed but arbitrary.

(c) At least for $|x| \le 1 - \delta$, $\delta \in (0,1)$ fixed but arbitrary, the term $n^{-1}\mathrm{Re}[...]$ in the right–hand side of (2.16) is $o(1)$, and even $O(Q(a_n)/n)$ – see Section 15. However it is not $O(n^{-\delta})$ in general.

(d) The density function $\mu_{n,a_n}(x)$ describes the distribution of the zeros of $p_n(W^2,x)$. For the Erdös case, one naturally has an "arcsine distribution":

$$\lim_{n\to\infty} \pi (1 - x^2)^{1/2} \mu_{n,a_n}(x) = 1,$$

uniformly for $|x| \le 1 - \delta$, any fixed $\delta \in (0,1)$. Lemma 4.4 contains a proof of this, and the proof there yields the rate $O(Q(a_n)/n)$.

(e) Analogues of Theorem 2.3 for Freud weights appear in the appendix.

(f) For $W \in SE$, one can prove a sharper form of (2.16) if we replace W^2 by $1/G^2_{Q/2}$ – see Theorem 15.1. We need $W \in SE(3)$ only in passing from $1/G^2_{Q/2}$ to W^2.

Next, we turn to orthogonal polynomials on $\mathbb{R}$:

<u>Theorem 2.4</u>

Let $W \in SE(3)$, with the additional restriction that $Q'(x)$ is continuous at $x = 0$. For $n \ge 1$, let c_n be defined by (2.11) with s fixed, but large enough, and let $\hat{F}_n$ be defined by (2.12) with $p = 2$. Then there exists $\delta > 0$, such that as $n \to \infty$, uniformly for

$$|x| \le 1 - n^{-\delta},\ x = \cos\phi,\ z := e^{i\phi}, \tag{2.24}$$

we have

$$c_n^{1/2} p_n(W^2, c_n x)\, \hat{F}_n(x)$$

$$= (2/\pi)^{1/2}\, \mathrm{Re}\, \{z^n\, \hat{F}_n(\cos\phi)\, /\, \overline{D(\hat{F}_n^2(\cos t); z)}\} + O(n^{-\delta}), \qquad (2.25)$$

$$= (2/\pi)^{1/2} \cos\,(n\phi + \Gamma(\hat{F}_n^2; \phi)) + O(n^{-\delta}). \qquad (2.26)$$

$$= (2/\pi)^{1/2} \cos\,(n\pi \int_x^1 \mu_{n,c_n}(t)\, dt + \phi/2 - \pi/4) + O(n^{-\delta}), \qquad (2.27)$$

where $\Gamma(\cdot\,;\cdot\,)$ is defined by (1.21), and μ_{n,c_n} is defined by (4.3) to (4.5) below.

<u>Proof</u>

This is the special case m = n of Theorem 13.4. □

<u>Remarks</u>

(a) The asymptotic above is possibly equivalent to that where s = 0 (so that $c_n = a_n$), but I have not been able to prove this.

(b) In Plancherel–Rotach type asymptotics, [1,22,29], one usually has the more restrictive range $|x| \leq 1 - \delta$, $\delta > 0$ arbitrary.

(c) One can multiply W by various factors that admit suitable relative polynomial approximations – see Theorems 13.1 and 13.2.

(d) There is an analogue of Theorem 2.3 for the weight $1/G_Q^2$, assuming only $W \in SE$ (that is, only two derivatives) see Theorem 13.7. Unfortunately, in passing from $G_{Q/2}^{-2}$ to W^2, via the techniques of the relativized Szegö theory [17,28], one needs good asymptotics for G_Q that require three derivatives (Theorem 6.1). When Q(x) grows so fast that, for example, (2.14) holds, we can weaken the smoothness conditions.

(e) For Freud weights, the analogue of Theorem 2.4 appears in the appendix.

Next, we present results on the largest zeros of extremal polynomials:

Theorem 2.5

Let $W \in SE$. Let j be a fixed positive integer. Then there exist C and n_1 such that

$$|x_{jn}^{(p)}/a_n - 1| \leq C\,\Delta_n, \tag{2.28}$$

uniformly for $2 \leq p \leq \infty$ and $n \geq n_1$.

Proof

See Theorem 7.1. □

Remarks

(a) Theorem 7.1 contains an upper bound for $x_{jn}^{(p)}$ for $0 < p < 1$.

(b) In the Freud case [14,Theorem 4.4,p.27], it is known that

$$x_{jn}^{(p)}/a_n - 1 = O\{(\log n)/n\}^{2/3},$$

and one expects that [18]

$$x_{jn}^{(p)}/a_n - 1 = Cn^{-2/3}\{1 + o(1)\},\ n \to \infty.$$

In contrast, (2.28) shows that for the Erdös case

$$x_{jn}^{(p)}/a_n - 1 = O\{(\log n)/(nA_n^*)\}^{2/3} = o\{(\log n)/n\}^{2/3},$$

and one expects that

$$x_{jn}^{(p)}/a_n - 1 = C(nA_n^*)^{-2/3}\{1 + o(1)\} = C(nT(a_n))^{-2/3}\{1 + o(1)\},\ n \to \infty.$$

More generally, one expects that for either Freud or Erdös weights,

$$x_{jn}^{(p)}/a_n - 1 = C(nA_{n,a_n})^{-2/3}\{1 + o(1)\},\ n \to \infty,$$

where A_{n,a_n} is defined by (4.12) or (4.17) below, and C may well be independent of W.

Finally, a Markov–Bernstein Inequality:

Theorem 2.6

Let $W \in SE$. Then for $n \geq 1$,

$$\sup_{P \in \mathcal{P}_n} \|P'W\|_{L_\infty(\mathbb{R})} / \|PW\|_{L_\infty(\mathbb{R})} \sim Q'(a_n) \sim (n/a_n) T(a_n)^{1/2}. \tag{2.29}$$

Proof

See Corollary 7.3. The difficult part – the upper bound in the ~ – was proved in [13,Thm.1.3]. □

Remark

In contrast to the Freud case [9,25], note that the dependence on n grows essentially faster than n/a_n, as $n \to \infty$. However, $T(a_n)$ grows slowly, and one expects no faster than $(\log n)^{1+\epsilon}$, $\epsilon > 0$ fixed, but arbitrary. The weak restriction (2.3) yields only

$$T(a_n) = O((n/a_n)^{2/29}), \quad n \to \infty.$$

There are several auxiliary results – notably Theorems 5.1, 5.2, 6.1, 7.1, 8.1, 9.1, 10.1 and Lemma 11.8 – that may be of independent interest to the reader. At a first reading of the proofs, it may be best to start with Sections 4 and 5, then Sections 7 to 15. Section 6 is independent of the rest of this paper, and proves results in greater generality, which we shall use in the appendix. Parts of the appendix can be read independently of the main part of this paper, but other parts refer back to similar proofs. Papers that the reader shall definitely need in wading through the proofs are [13], as well as the books [4,14,30].

3 Preliminaries

In this section, we list some preliminary technical lemmas.

Lemma 3.1

Let $W \in SE$, except that we do not assume (2.3) or (2.4).

(a) Given $r > 0$, there exists C such that

$$Q^{(j)}(x) \geq x^r,\ x \geq C,\ j = 0, 1, 2. \tag{3.1}$$

(b) Given $r > 0$, $Q''(x)/x^r$ and $Q'(x)/x^r$ are increasing for x large enough.

(c) For $L \geq 1$ and $x \in (0, \infty)$,

$$L^{T(x)-1} \leq Q'(Lx)/Q'(x) \leq L^{T(Lx)-1}. \tag{3.2}$$

(d)

$$\limsup_{x \to 0+} x^{1-T(0+)}\, Q'(x) < \infty, \tag{3.3}$$

and

$$\limsup_{x \to 0+} x^{2-T(0+)} |Q''(x)| < \infty. \tag{3.4}$$

(e) For $j = 0, 1, 2$ and each fixed $L > 1$,

$$\lim_{x \to \infty} Q^{(j)}(Lx)/Q^{(j)}(x) = \infty. \tag{3.5}$$

(f) For $j = 0, 1$,

$$\lim_{x \to \infty} xQ^{(j+1)}(x)/Q^{(j)}(x) = \infty. \tag{3.6}$$

(g) Given $r > 1$, there exist C_1 and C_2 such that

$$T(x) \leq C_1 + C_2 \log \{Q'(rx)/Q'(x)\},\ x \in (0, \infty). \tag{3.7}$$

(h) If also, Q'' is continuous in $\mathbb{R}$, then there exist C and $s > 0$ such that

$$Q'(x)/x \le C\, Q'(y)/y,\ 0 < x \le y,\ y \ge s, \tag{3.8}$$

and

$$|Q^{(j)}(x)| \le C\,|Q^{(j)}(y)|,\ 0 < x \le y;\ y \ge s;\ j = 1,2. \tag{3.9}$$

(i) There exists C_1, C_2 such that

$$Q'(x)/Q(x) \le C_1\, Q''(x)/Q'(x) \sim T(x)/x,\ x \ge C_2. \tag{3.10}$$

(j) If in addition, $W \in SE(3)$, then there exists C such that

$$Q'''(x)/Q''(x) \sim T(x)/x,\ x \ge C. \tag{3.11}$$

Proof

(a), (e), (f), (g), (h) are respectively Lemma 2.1 (a), (e), (f), (g), (h) in [13].

(b) Now

$$\frac{d}{dx}\{Q'(x)/x^r\} = \{Q'(x)/x^{r+1}\}\,\{\frac{xQ''(x)}{Q'(x)} - r\} > 0,\ x \text{ large enough},$$

by (f) of this lemma. Further, the identity (see (1.13))

$$Q''(x)/x^r = \{Q'(x)/x^{r+1}\}\,\{T(x) - 1\}$$

shows that $Q''(x)/x^r$ is the product of two increasing functions of x, for x large enough.

(c) For $L \ge 1$ and $x \in (0, \infty)$,

$$\frac{LxQ'(Lx)}{xQ'(x)} = \exp(\int_x^{Lx} \frac{(tQ'(t))'}{tQ'(t)} dt\,) = \exp(\int_x^{Lx} \frac{T(t)}{t}\, dt),$$

and the monotonicity of T yields that

$$T(x) \log L \le \int_x^{Lx} \frac{T(t)}{t}\, dt \le T(Lx) \log L.$$

(d) Applying (c) of this lemma yields for $x \in (0,1)$,

$$xQ'(x) \le Q'(1)\, x^{T(x)} \le Q'(1)\, x^{T(0+)}.$$

Then (3.3) follows, while (3.4) follows from (3.3) and the identity

$$xQ''(x) = Q'(x)\,\{T(x) - 1\}.$$

(i) Since $T(x) \to \infty$ as $x \to \infty$, the $\sim$ in (3.10) follows. Next, for x large enough,

$$Q(x) = Q(x/2) + xQ'(x) \int_{1/2}^{1} Q'(ux)/Q'(x)\, du$$

$$\geq 0 + xQ'(x) \int_{1/2}^{1} u^{T(x)-1}\, du \text{ (by (c) of this lemma)}$$
$$= xQ'(x)\, \{1 - (1/2)^{T(x)}\}/T(x) \geq C\, xQ'(x)/T(x),$$

since $T(x) \geq T(0+) > 0$. This yields the left half of (3.10).

(j) Since $T(x)$ is increasing,

$$0 \leq T'(x) = \frac{Q''(x)}{Q'(x)} + \frac{xQ'''(x)}{Q'(x)} - x \left(\frac{Q''(x)}{Q'(x)}\right)^2.$$

Multiplying by $Q'(x)/(xQ''(x))$ yields

$$\frac{Q'''(x)}{Q''(x)} \geq \frac{Q''(x)}{Q'(x)} - \frac{1}{x} \geq \{C_1 T(x) - 1\}/x \geq C_2 T(x)/x,$$

for x large enough, by (3.10). To obtain a corresponding upper bound, one uses (2.4) and (2.5). □

Lemma 3.2

Let $W \in SE$, except that we do not assume (2.3) or (2.4).

(a) Then

$$\lim_{n\to\infty} a_n^j\, Q^{(j)}(a_n)/n = \begin{cases} 0, & j = 0, \\ \infty, & j = 1,2. \end{cases} \tag{3.12}$$

(b) Uniformly for x in compact subsets of (0,1), we have

$$\lim_{n\to\infty} a_n^j\, Q^{(j)}(a_n x)/n = 0,\ j = 0,1,2. \tag{3.13}$$

(c) For $j = 1,2$ and n large enough,

$$a_n^j\, Q^{(j)}(a_n)/n \leq C\, T(a_n)^{j-1/2}. \tag{3.14}$$

If in addition, (2.4) holds, then for $j = 0, 1, 2$,

$$a_n^j\, Q^{(j)}(a_n)/n \sim T(a_n)^{j-1/2}. \tag{3.15}$$

(d) There exist C_1 and C_2 such that

$$(C_1 u\, T(a_u))^{-1} \leq a_u'/a_u \leq (C_2 u\, T(a_u/2))^{-1},\ u \in [0, \infty). \tag{3.16}$$

(e) There exists C such that

$$a_{ru}/a_u \geq 1 + C\, (\log r)/\, T(a_{ru}),\ r \in [1, \infty),\ u \in [0, \infty). \tag{3.17}$$

(f) For each fixed $L > 0$,

$$\lim_{u\to\infty} a_{Lu}/a_u = 1. \tag{3.18}$$

(g) For each fixed $\delta > 0$, a_n/n^δ is decreasing for large enough n, and

$$\lim_{n\to\infty} a_n\, n^{-\delta} = 0. \tag{3.19}$$

(h) For $m \geq n \geq 1$,

$$a_m/a_n \leq (m/n)^{1/T(0+)}. \tag{3.20}$$

(i) Given $\epsilon > 0$, there exists n_0 such that

$$a_m/a_n \leq (m/n)^\epsilon,\ m \geq n \geq n_0, \tag{3.21}$$

and

$$|a_m/a_n - 1| \leq \epsilon\, |m/n - 1|,\ n \geq n_0,\ n/2 \leq m \leq 2n. \tag{3.22}$$

(j) A_u^* (defined by (1.11)) is an increasing function of $u \in (0, \infty)$, with

$$\lim_{u\to\infty} A_u^* = \infty.$$

<u>Proof</u>

(a) – (g) are Lemma 2.2 (a) – (g) in [13], except for (3.15) and the decrescence of a_n/n^δ. However, the latter was proved in [13, proof of Lemma 2.2(g)]. We turn to the proof of (3.15). Let $r := r(n) := 1 - 1/T(a_n)$, n large enough. We have by definition of a_n,

$$\frac{n}{a_n Q'(a_n)} = \frac{2}{\pi}\left[\int_0^r + \int_r^1\right] \frac{a_n t Q'(a_n t)}{a_n Q'(a_n)} \frac{dt}{(1-t^2)^{1/2}}$$

$$\leq \frac{2}{\pi}(1-r^2)^{-1/2}\{a_n Q'(a_n)\}^{-1}\int_0^r a_n Q'(a_n t)\, dt + \frac{2}{\pi}\int_r^1 (1-t^2)^{-1/2}\, dt$$

$$\leq C\, T(a_n)^{1/2}\{a_n Q'(a_n)\}^{-1} Q(a_n) + C\, T(a_n)^{-1/2} \leq C_1\, T(a_n)^{-1/2},$$

by (2.4) and (3.10). Together with (3.14), this establishes (3.15) for $j = 1$. Since (2.4) and (3.10) show that for x large enough,

$$x^2 Q''(x) \sim xQ'(x)\, T(x),$$

and

$$Q(x) \sim xQ'(x)/T(x),$$

(3.15) follows also for j = 0, 2.

(h) Differentiating (1.10) with respect to u, yields

$$1 = \frac{a'_u}{a_u} \frac{2}{\pi} \int_0^1 a_u t Q'(a_u t)\, T(a_u t)\, (1 - t^2)^{-1/2}\, dt \tag{3.23}$$
$$\geq \frac{a'_u}{a_u} T(0+)\, u, \quad u \in (0, \infty).$$

Then for $m \geq n \geq 1$,

$$a_m/a_n = \exp\left(\int_n^m a'_u/a_u\, du \right) \leq \exp\left(T(0+)^{-1} \int_n^m u^{-1}\, du \right) = (m/n)^{T(0+)}.$$

(i) From (3.16), for u large enough,

$$a'_u/a_u \leq \epsilon/u.$$

Integrating as before, yields (3.21). Then (3.22) follows easily, by considering separately the cases $m < n$ and $m \geq n$.

(j) Since $u^2 Q''(u) = uQ'(u)[T(u) - 1]$, and $uQ'(u)$, $T(u)$ are increasing in $(0, \infty)$, the monotonicity of A^*_u follows from (1.11). Further, taking into account (3.6) and (1.10), we obtain $\lim_{u\to\infty} A^*_u = \infty$. □

Next, one more lemma on a_n:

Lemma 3.3

Let $W \in SE$, except that we do not assume (2.4).

(a) Then as $n \to \infty$,

$$Q'(a_n) = O((n/a_n)^{30/29}), \tag{3.24}$$
$$T(a_n) = O((n/a_n)^{2/29}), \tag{3.25}$$

and

$$a_n Q''(a_n) = O((n/a_n)^{32/29}). \tag{3.26}$$

(b) Suppose

$$m = m(n) = n[1 + O((n/a_n)^{-2/29})],\ n \to \infty. \tag{3.27}$$

Then

$$\lim_{n\to\infty} Q'(a_m)/Q'(a_n) = 1. \tag{3.28}$$

(c) Suppose

$$x = x(n) = a_n[1 + o((n/a_n)^{-2/29})],\ n \to \infty. \tag{3.29}$$

Then as $n \to \infty$,

$$Q'(x) = O((n/a_n)^{30/29}), \tag{3.30}$$

and

$$a_n Q''(x) = O((n/a_n)^{32/29}). \tag{3.31}$$

(d) If $\{\epsilon_n\}_{n=1}^{\infty} \subset [0, \infty)$ satisfies

$$\epsilon_n = O(n^{-2/29}),\ n \to \infty, \tag{3.32}$$

then

$$a_n(1 + \epsilon_n) \leq a_{2n},\ \text{n large enough}. \tag{3.33}$$

In particular, for each fixed $s \geq 0$,

$$a_n(1 + s\,\Delta_n) \leq a_{2n},\ \text{n large enough}. \tag{3.34}$$

(e)

$$n^{27/29} = o(Q(a_n)),\ n \to \infty. \tag{3.35}$$

(f) Suppose that $\{e_n\}_{n=1}^{\infty} \subset (0, \infty)$ satisfies

$$e_n/a_n = 1 + \delta_n\,(n\,T(a_{2n}))^{-1/2},\ n \to \infty, \tag{3.36}$$

where

$$\lim_{n\to\infty} \delta_n = 0. \tag{3.37}$$

Then

$$e_n^n\, G[W(e_n x)] \,/\, \{a_n^n\, G[W(a_n x)]\} = 1 + O(\delta_n^2),\ n \to \infty. \tag{3.38}$$

<u>**Proof**</u>

(a) – (c) are the special case $\eta = 1/30$ of Lemma 2.3 in [13].

(d) From Lemma 3.2(e), and from (3.25),

$$a_{2n}/a_n \geq 1 + C \log 2/T(a_{2n}) \geq 1 + C_1 (a_{2n}/n)^{2/29}.$$

Then (3.32) yields (3.33). Next, recall from Lemma 3.2(j) that

$$\lim_{n\to\infty} A_n^* = \infty. \tag{3.39}$$

Hence, from (1.12),

$$\Delta_n = o((\log n)/n)^{2/3} = O(n^{-2/29}), \ n \to \infty, \tag{3.40}$$

so (3.34) follows.

(e) By the monotonicity of Q'(t) for large t, we have for large n,

$$Q(a_n) = Q(a_{n/2}) + \int_{a_{n/2}}^{a_n} Q'(t)\,dt$$

$$\geq 0 + a_{n/2}Q'(a_{n/2})\,(a_n/a_{n/2} - 1)$$

$$\geq C\, a_{n/2}Q'(a_{n/2})/T(a_n) \text{ (by Lemma 3.2(e))}$$

$$\geq n\,(a_n/n)^{2/29},$$

for n large enough, by Lemma 3.2(a) and (3.25).

(f) Now expanding about a_n,

$$\log \{ e_n^n\, G[W(e_n x)] \} = n \log e_n - \frac{2}{\pi}\int_0^1 Q(e_n t)\,(1-t^2)^{-1/2}\,dt$$

$$= n \log a_n - \frac{2}{\pi}\int_0^1 Q(a_n t)\,(1-t^2)^{-1/2}\,dt$$

$$+ (e_n - a_n)\,\{ \frac{n}{a_n} - \frac{2}{\pi}\int_0^1 t\,Q'(a_n t)\,(1-t^2)^{-1/2}\,dt \}$$

$$+ \frac{1}{2}(e_n - a_n)^2\,\{ -\frac{n}{\xi^2} - \frac{2}{\pi}\int_0^1 t^2\,Q''(\xi t)\,(1-t^2)^{-1/2}\,dt \},$$

where $\xi = \xi(n)$ lies between e_n and a_n. Note that by definition of a_n, the second term in this last right–hand side, is zero. Further, the differentiation under the integral sign is justified in view of Lemma 3.1(d). As $\xi \sim a_n$, we deduce that

$$| \log \left(\{e_n^n \, G[W(e_n x)]\} / \{a_n^n \, G[W(a_n x)]\} \right) |$$
$$\leq C \, (e_n/a_n - 1)^2 \, n \, \{1 + \frac{2}{n\pi} \int_0^1 (\xi t)^2 \, | Q''(\xi t)| \, (1 - t^2)^{-1/2} \, dt\}. \tag{3.41}$$

Now from (d) of this lemma, and from (3.36),

$$\xi \leq \max\{e_n, a_n\} \leq a_{2n}, \text{ n large enough.}$$

Next, as $T(t)$ and $tQ'(t)$ are increasing in $(0,\infty)$,

$$(\xi t)^2 \, | Q''(\xi t)| = \xi t \, Q'(\xi t) \, | T(\xi t) - 1|$$
$$\leq a_{2n} t \, Q'(a_{2n} t) \, T(a_{2n}),$$

so from (1.10),

$$\frac{2}{\pi} \int_0^1 (\xi t)^2 \, | Q''(\xi t)| \, (1 - t^2)^{1/2} \, dt \leq T(a_{2n}) \, 2n,$$

Then, from (3.36), we can bound above the right hand side of (3.41) for n large enough, by

$$O(\delta_n^2 \, T(a_{2n})^{-1}) \, \{1 + C \, T(a_{2n})\} = O(\delta_n^2). \qquad \square$$

<u>Lemma 3.4</u>

Let $W \in SE(3)$.

(a) There exists C such that for t large enough,

$$| T'(t)| \leq C \, T^2(t)/t. \tag{3.42}$$

(b) For t large enough,

$$a_t'/a_t \sim (tT(a_t))^{-1}. \tag{3.43}$$

(c) For u large enough,

$$A_{2u}^* \sim T(a_{2u}) \sim T(a_u) \sim A_u^*. \tag{3.44}$$

(d) There exist C and n_1 such that for $n \geq n_1$ and $n/2 \leq m \leq 2n$,

$$| T(a_m)/T(a_n) - 1| \leq C \, | m/n - 1| , \tag{3.45}$$

$$| A_m^*/A_n^* - 1| \leq C \, | m/n - 1|^{1/2}, \tag{3.46}$$

and

$$|\Delta_m/\Delta_n - 1| \le C\,|m/n - 1|^{1/2}. \tag{3.47}$$

<u>Proof</u>

(a) Since

$$T'(t) = \frac{tQ'''(t)}{Q'(t)} + \frac{Q''(t)}{Q'(t)} - t\,\left(\frac{Q''(t)}{Q'(t)}\right)^2,$$

for t large enough, (2.4), (2.5) (with k = 3) and (3.10) yield (3.42).

(b) In view of Lemma 3.2(d), it suffices to show that

$$a'_u/a_u \le C/(uT(a_u)),\ u \text{ large enough}. \tag{3.48}$$

Let $r := r(u) := 1 - T(a_u)^{-1}$, u large enough. Now from (3.23) and Lemma 3.1(c),

$$1 \ge \frac{2}{\pi}\frac{a'_u}{a_u}\int_0^1 t^{T(a_u)}\, a_uQ'(a_u)\, T(a_ut)\,(1-t^2)^{-1/2}\,dt$$

$$\ge \frac{2}{\pi}\frac{a'_u}{a_u}\, r^{T(a_u)}\, a_uQ'(a_u)\, T(a_ur)\int_r^1 (1-t^2)^{-1/2}\,dt$$

$$\ge C\,\frac{a'_u}{a_u}\, a_uQ'(a_u)\, T(a_ur)\, T(a_u)^{-1/2},$$

by choice of r. Here, using (a) of this lemma,

$$1 \le T(a_u)/T(a_ur) = \exp\left(\int_{a_ur}^{a_u} T'(t)/T(t)\,dt\right)$$

$$\le \exp\left(C\int_{a_ur}^{a_u} T(t)/t\,dt\right)$$

$$\le \exp\left(T(a_u)\,C\log r^{-1}\right) \le C_1,$$

by choice of r. Thus

$$1 \ge C_2\,\frac{a'_u}{a_u}\, a_uQ'(a_u)\, T(a_u)^{1/2},$$

and then (3.15) (with j = 1) yields (3.48).

(c) Firstly, for u large enough,

$$A_u^* = \frac{2}{u\pi}\int_0^1 a_u t\ Q'(a_u t)\ \{T(a_u t) - 1\}\ (1-t^2)^{-1/2}\ dt$$

$$\leq T(a_u) - 1 \leq T(a_u), \tag{3.49}$$

by (1.10). Next, if $K > 0$ is fixed, a substitution shows that

$$\int_0^{K/a_u} (a_u t)^2\ Q''(a_u t)\ (1-t^2)^{-1/2}\ dt = O(1/a_u),\ u \to \infty.$$

Let us choose K so large that $T(s) \geq 2$, $s \geq K$. Then, with r as in (b) above,

$$\int_{K/a_u}^1 (a_u t)^2\ Q''(a_u t)\ (1-t^2)^{-1/2}\ dt$$

$$= \int_{K/a_u}^1 a_u t Q'(a_u t)\ \{T(a_u t) - 1\}\ (1-t^2)^{-1/2}\ dt$$

$$\geq (1/2)\int_{K/a_u}^1 a_u t Q'(a_u t)\ T(a_u t)\ (1-t^2)^{-1/2}\ dt$$

$$\geq (1/2)\int_r^1 t^{T(a_u)}\ a_u Q'(a_u)\ T(a_u t)\ (1-t^2)^{-1/2}\ dt$$

$$\geq C\ a_u Q'(a_u)\ T(a_u)^{1/2} \geq C_1\ u\ T(a_u),$$

exactly as in (b). Together with (3.49), this establishes that

$$A_u^* \sim T(a_u),\ u \text{ large enough.} \tag{3.50}$$

Next,

$$1 \leq T(a_{2u})/T(a_u) = \exp\left(\int_u^{2u} \frac{T'(a_t)a_t'}{T(a_t)}\ dt\right)$$

$$\leq \exp\left(C_2 \int_u^{2u} t^{-1}\ dt\right) = 2^{C_2},$$

by (3.42) and (3.43). Together with (3.50), this yields (3.44).

(d) Firstly, (3.45) follows using the same argument as after (3.50). Also, (3.47) follows from (3.46) and (1.12). It remains to prove (3.46). Recall from Lemma 3.2(j) that A_u^* is an increasing function of u. Suppose that $n \geq m \geq n/2$, and n is large enough. Then, making a substitution in the integral (1.11) defining A_m^*,

$$A_m^* = \frac{2}{m\pi}\int_0^{a_m/a_n} (a_n t)^2\, Q''(a_n t)\, (1-(a_n t/a_m)^2)^{-1/2}\, \{a_n/a_m\}\, dt$$

$$\geq \frac{n}{m}\frac{a_n}{a_m}\frac{2}{n\pi}\int_0^{a_m/a_n} (a_n t)^2\, Q''(a_n t)\, (1-t^2)^{-1/2}\, dt$$

$$\geq \{A_n^* - \frac{2}{n\pi}\int_{a_m/a_n}^{1} (a_n t)^2\, Q''(a_n t)\, (1-t^2)^{-1/2}\, dt\}$$

$$\geq \{A_n^* - C\, a_n^2 Q''(a_n)\, n^{-1}\, (1-a_m/a_n)^{1/2}\}$$

$$\geq A_n^*\, \{1 - C_1 T(a_n)^{3/2} T(a_n)^{-1} (1-a_m/a_n)^{1/2}\}, \qquad (3.51)$$

by (3.15) and by (c) of this lemma. Finally, from (3.43),

$$1 \leq a_n/a_m \leq \exp(\, C\int_m^n (tT(a_t))^{-1}\, dt\,)$$

$$\leq \exp\,(C\, T(a_m)^{-1} \log(n/m)\,)$$

$$\leq 1 + C_2\, T(a_m)^{-1} \log\,(n/m)$$

$$\leq 1 + C_3\, T(a_m)^{-1}\, \{n/m - 1\},$$

so

$$(1-a_m/a_n)^{1/2} \leq C_4\, T(a_m)^{-1/2}\, (n/m-1)^{1/2}.$$

Then (3.46) follows. □

4 Potentials and Majorisation Functions

In this section, we gather together some results on integral equations with logarithmic kernel, and on potentials and majorisation functions, many of which were proved in [14], and to a lesser extent, in [13].

Lemma 4.1

Let $W := e^{-Q}$, where $Q(x)$ is continuous in $\mathbb{R}$, $Q''(x)$ exists and is continuous in $(0,\infty)$, while $Q'(x)$ is positive there, and $xQ'(x)$ is strictly increasing in $(0,\infty)$, with

$$\lim_{x\to\infty} xQ'(x) = \infty, \tag{4.1}$$

and for some $\eta \in (0,1)$,

$$\lim_{x\to 0+} x^{\eta}Q'(x) = 0. \tag{4.2}$$

(a) Let $n \geq 1$ and $0 < R \leq a_n$. Define for a.e. $x \in (-1,1)$,

$$\mu_{n,R}(x) := \nu_{n,R}(x) + B_{n,R}\, \pi^{-1} (1 - x^2)^{-1/2}, \tag{4.3}$$

where

$$\nu_{n,R}(x) := 2\pi^{-2} \int_0^1 \frac{(1 - x^2)^{1/2}}{(1 - s^2)^{1/2}} \frac{(RsQ'(Rs) - RxQ'(Rx))}{n(s^2 - x^2)}\, ds, \tag{4.4}$$

and

$$B_{n,R} := 1 - 2\,(n\pi)^{-1} \int_0^1 \frac{RtQ'(Rt)}{(1 - t^2)^{1/2}}\, dt. \tag{4.5}$$

Then

$$\mu_{n,R}(x) \geq \nu_{n,R}(x) > 0, \text{ a.e. } x \in (-1,1), \tag{4.6}$$

and

$$\int_{-1}^{1} \mu_{n,R}(x)\, dx = 1, \tag{4.7}$$

and for some $1 < p < 2$,

$$\|\mu_{n,R}\|_{L_p[-1,1]} \leq C_1 \, \|RtQ'(Rt)n^{-1}(1 - t^2)^{-1/2}\|_{L_p[-1,1]} + C_2 \, B_{n,R}, \tag{4.8}$$

where C_1 and C_2 are independent of n,Q and R. Further,

$$0 \leq B_{n,R} \leq 1, \tag{4.9}$$

and

$$B_{n,R} = 0 \text{ iff } R = a_n, \tag{4.10}$$

and for $x \in [-1,1]$,

$$\int_{-1}^{1} \log |x - t| \; \mu_{n,R}(t)\, dt$$

$$= Q(Rx)/n - 2\, (n\pi)^{-1} \int_0^1 Q(Rt)\, (1 - t^2)^{-1/2}\, dt - \log 2. \tag{4.11}$$

(b) Define for $0 < R \leq a_n$,

$$A_{n,R} := 2\, (n\pi^2)^{-1} \int_0^1 \frac{RQ'(R) - RtQ'(Rt)}{(1 - t^2)^{3/2}}\, dt, \tag{4.12}$$

$$\tau_{n,R} := RQ'(R)/n + \max\{|R^2Q''(Ru)|/n \colon u \in [1/2,1]\}, \tag{4.13}$$

and given $\epsilon > 0$,

$$\rho_{n,R,\epsilon} := \max\{|R^2Q''(Ru)|/n \colon u \in [1,1 + \epsilon]\}. \tag{4.14}$$

Then for some C_3 independent of n,R and Q,

$$|\nu_{n,R}(x)\, (1 - x^2)^{-1/2} - A_{n,R}| \leq C_3\, (1 - x)^{1/5}\, \tau_{n,R}, \tag{4.15}$$

for $x \in [7/8,1)$ and $0 < R \leq a_n$. Further,

$$\int_{-1}^{1} \nu_{n,R}(x)\, (1 - x)^{-1}\, dx = RQ'(R)/n. \tag{4.16}$$

and (' denotes differentiation with respect to t)

$$A_{n,R} = 2\, (n\pi^2)^{-1} \int_0^1 \frac{t(RtQ'(Rt))'}{(1 - t^2)^{1/2}}\, dt\, . \tag{4.17}$$

(c) For $z \in \mathbb{C}$, let

$$U_{n,R}(z) := \int_{-1}^{1} \log|z-t| \, \mu_{n,R}(t)\, dt - Q(R|z|)/n + \chi_{n,R}/n, \tag{4.18}$$

where

$$\chi_{n,R} := 2\pi^{-1} \int_{0}^{1} \frac{Q(Rt)}{(1-t^2)^{1/2}}\, dt + n \log 2. \tag{4.19}$$

Then

$$U_{n,R}(x) = 0, \; x \in [-1,1], \tag{4.20}$$

and as $\epsilon \to 0+$, uniformly for $0 < R \le a_n$, $n \ge 1$,

$$U'_{n,R}(1+\epsilon)$$
$$= -A_{n,R}\, \pi(2\epsilon)^{1/2} + B_{n,R}((1+\epsilon)^2-1)^{-1/2} + O(\tau_{n,R}\, \epsilon^{2/3}) + O(\rho_{n,R,\epsilon}\, \epsilon), \tag{4.21}$$

and

$$U_{n,R}(1+\epsilon)$$
$$= -A_{n,R}\, \pi\sqrt{8}\epsilon^{3/2}/3 + B_{n,R} \log \varphi(1+\epsilon) + O(\tau_{n,R}\, \epsilon^{5/3}) + O(\rho_{n,R,\epsilon}\, \epsilon^2), \tag{4.22}$$

where the constants in the order terms are independent of n, ϵ, R and Q. Finally, for $0 < R \le a_n$ and $n \ge 1$,

$$(xU'_{n,R}(x))' < 0, \; x \in (1, \infty). \tag{4.23}$$

If $R = a_n$, then also

$$U_{n,R}(x) < 0 \; ; \; U'_{n,R}(x) < 0, \; x > 1. \tag{4.24}$$

<u>Proof</u>

This is Lemma 5.3 in [14,pp.37–39], with some very minor changes. Note that (4.2) implies that for $1 < p < 1/\eta$,

$$\|Q'\|_{L_p[0,1]} < \infty,$$

so (5.36) in [14] is satisfied. Further (4.20) is a restatement of (4.11), and (4.17) may be derived from (4.12), with the aid of an integration by parts. □

We shall need several estimates for the constants and functions appearing in Lemma 4.1.

<u>Lemma 4.2</u>

Let $W \in SE$, except that we do not assume (2.3) or (2.4). Assume the notation of Lemma 4.1, and define for $0 < R \leq a_n$ and $n \geq 1$,

$$A^*_{n,R} := \frac{2}{n\pi} \int_0^1 (Rt)^2\, Q''(Rt)\, (1 - t^2)^{-1/2}\, dt. \tag{4.25}$$

(a) Then $A^*_{n,R}$ is an increasing function of $R \in (0,a_n]$, and,

$$A^*_{n,R} \leq T(a_n), \tag{4.26}$$

and

$$A^*_{n,R} \leq B_{n,R}/(1 - R/a_n) \leq (a_n/R)\,\{1 + A^*_n\}. \tag{4.27}$$

(b) If $\tau_{n,R}$ and $\rho_{n,R,\epsilon}$ are defined by (4.13–14), then for $n \geq n_0$, and $a_n/2 \leq R \leq a_n$,

$$\tau_{n,R} \leq 2R^2\, Q''(R)/n \leq C_1\, T(a_n)^{3/2}, \tag{4.28}$$

and for some C_1 independent of n, R, ϵ,

$$\rho_{n,R,\epsilon} \leq R^2 Q''(R(1 + \epsilon))/n \leq C_1\, T(a_n(1 + \epsilon))^{3/2}\, (1 + \epsilon)^{T(a_n(1 + \epsilon))} \tag{4.29}$$

(c) For $n \geq n_0$, and $a_n/2 \leq R \leq a_n$,

$$A^*_{n,R} \leq A^*_n \leq A^*_{n,R} + C_1\, T(a_n)^{3/2}\, (1 - R/a_n)^{1/2}. \tag{4.30}$$

(d) For $0 < R \leq a_n$,

$$A^*_{n,R}/\pi \leq A_{n,R} \leq \{A^*_{n,R} + 1\}/\pi, \tag{4.31}$$

and uniformly for $R \in [a_n/2,a_n]$,

$$\lim_{n\to\infty} A^*_{n,R}/A_{n,R} = \pi. \tag{4.32}$$

Proof

(a) Since $t^2Q''(t) = (T(t) - 1)\, tQ'(t)$ is an increasing function of t, the monotonicity of $A^*_{n,R}$ follows. The inequality (4.26) then follows as in Lemma 3.4 – see (3.49). Next, from (4.5) and (1.10),

$$B_{n,R} = \frac{2}{n\pi}\int_0^1 \{a_n tQ'(a_n t) - RtQ'(Rt)\}\ (1 - t^2)^{-1/2}\ dt$$

$$= \frac{2}{n\pi}\int_0^1 \{a_n t - Rt\}\ (uQ'(u))'\ (1 - t^2)^{-1/2}\ dt,$$

where $u = u(n,t,R)$ lies between $a_n t$ and Rt. Then, as $(uQ'(u))'u = T(u)\ u\ Q'(u)$ is increasing for $u \in (0, \infty)$, we obtain

$$B_{n,R} \le \frac{2}{n\pi}\int_0^1 \{a_n t - Rt\}\ T(a_n t)\ a_n tQ'(a_n t)\ (Rt)^{-1}\ (1 - t^2)^{-1/2}\ dt$$

$$= (\frac{a_n}{R} - 1)[\frac{2}{n\pi}\int_0^1 a_n tQ'(a_n t)(1-t^2)^{-1/2}\ dt + \frac{2}{n\pi}\int_0^1 (a_n t)^2 Q''(a_n t)(1-t^2)^{-1/2}\ dt]$$

$$= \frac{a_n}{R}\ (1 - \frac{R}{a_n})\ [1 + A^*_n],$$

where we have used (1.13). Hence the upper bound in (4.27). For the lower bound, we obtain similarly

$$B_{n,R} \ge \frac{2}{n\pi}\int_0^1 \{a_n t - Rt\}\ T(Rt)\ RtQ'(Rt)\ (a_n t)^{-1}\ (1 - t^2)^{-1/2}\ dt$$

$$= (1 - \frac{R}{a_n})[\frac{2}{n\pi}\int_0^1 RtQ'(Rt)(1-t^2)^{-1/2}\ dt + \frac{2}{n\pi}\int_0^1 (Rt)^2 Q''(Rt)(1-t^2)^{-1/2}\ dt]$$

$$\ge (1 - \frac{R}{a_n})\ A^*_{n,R},$$

since $sQ'(s)$ is positive in $(0,\infty)$.

(b) From (4.13), for $n \ge n_0$, and $a_n/2 \le R \le a_n$, the monotonicity of Q'' yields

$$\tau_{n,R} \le n^{-1}\ [RQ'(R) + R^2Q''(R)]$$

$$\le 2n^{-1}\ a_n^2\ Q''(a_n) \le C\ T(a_n)^{3/2},$$

by (3.6) and (3.14). Next, from (4.14), for $n \ge n_0$ and $a_n/2 \le R \le a_n$,

$$\rho_{n,R,\epsilon} \le R^2\, Q''(R(1+\epsilon))/n$$

$$= (1+\epsilon)^{-2}\, [T(R(1+\epsilon)) - 1]\, R(1+\epsilon)\, Q'(R(1+\epsilon))/n$$

$$\le T(a_n(1+\epsilon))\, RQ'(R)\, (1+\epsilon)^{T(R(1+\epsilon))}/n$$

(by Lemma 3.1(c))

$$\le T(a_n(1+\epsilon))\, a_n Q'(a_n)\, (1+\epsilon)^{T(a_n(1+\epsilon))}/n$$

$$\le C\, T(a_n(1+\epsilon))^{3/2}\, (1+\epsilon)^{T(a_n(1+\epsilon))},$$

by (3.14).

(c) We need establish only the second inequality in (4.30). Now

$$A_n^* = \frac{2}{n\pi}\left[\int_0^{R/a_n} + \int_{R/a_n}^1\right] (a_n t)^2\, Q''(a_n t)\, (1-t^2)^{-1/2}\, dt$$

$$\le \frac{2}{n\pi}\int_0^1 (Rs)^2 Q''(Rs)(1-(Rs/a_n)^2)^{-1/2}\, \frac{R}{a_n}\, ds$$

$$+ a_n^2 Q''(a_n)\, \frac{2}{n\pi}\int_{R/a_n}^1 (1-t^2)^{-1/2} dt$$

$$\le \frac{R}{a_n}\, A_{n,R}^* + C\, T(a_n)^{3/2}\, (1-R/a_n)^{1/2},$$

by (3.14).

(d) From (4.17),

$$A_{n,R} = \frac{2}{n\pi^2}\left[\int_0^1 RtQ'(Rt)(1-t^2)^{-1/2}dt + \int_0^1 (Rt)^2 Q''(Rt)(1-t^2)^{-1/2}dt\right]. \tag{4.33}$$

Since $sQ'(s)$ is positive in $(0,\infty)$, we then obtain (4.31). Next, from (3.6), (with $j = 1$) we obtain for any $\{\xi_n\}_{n=1}^{\infty} \subset (0,\infty)$ with $\lim_{n\to\infty} \xi_n = \infty$, and uniformly for $R \in [a_n/2, a_n]$,

$$\lim_{n\to\infty} \int_{\xi_n/R}^{1} (Rt)^2Q''(Rt)(1-t^2)^{-1/2}\,dt \Big/ \int_{\xi_n/R}^{1} RtQ'(Rt)(1-t^2)^{-1/2}\,dt = \infty.$$

The corresponding integrals over the range $[0,\xi_n/R]$ will be relatively negligible if $\xi_n \to \infty$ sufficiently slowly. Hence (4.32). □

<u>Lemma 4.3</u>

Let $W \in SE$, except that we do not assume (2.3) or (2.4). Assume the notation of Lemma 4.1, and define for $0 < R \le a_n$ and n large enough,

$$\psi_{n,R}(x) := \int_{K/a_n}^{1} (1-s)^{-1/2} \frac{RxQ'(Rx) - RsQ'(Rs)}{Rx - Rs}\,ds, \tag{4.34}$$

where K is a fixed positive number, so large that $Q'(t)$ is increasing in $[K,\infty)$ and $T(t) \ge 2$, $t \in [K,\infty)$. Let $0 < \epsilon < 1$.

(a) $\psi_{n,R}(x)$ is an increasing function of $x \in [K/R,1]$.

(b) There exists n_1 such that

$$\nu_{n,R}(x) \sim \frac{R}{n}\,\psi_{n,R}(x)\,(1-|x|)^{1/2}, \tag{4.35}$$

uniformly for $\epsilon \le |x| \le 1$, $n \ge n_1$, and $a_n/2 \le R \le a_n$.

(c) There exist C and n_1 such that

$$\mu_{n,R}(x) \sim 1, \tag{4.36}$$

uniformly for $C/R \le |x| \le 1-\epsilon$, $n \ge n_1$, and $a_n/2 \le R \le a_n$. If $T(0+) > 1$, then (4.36) holds uniformly for $|x| \le 1-\epsilon$, $n \ge n_1$, and $a_n/2 \le R \le a_n$.

(d) There exists n_1 and C such that

$$\nu_{n,R}(x) \le C\,(1-B_{n,R})\,\{1 + \int_x^1 t^{T(0+)-2}\,dt\}, \tag{4.37}$$

uniformly for $x \in (0,1-\epsilon)$, $n \ge n_1$, and $a_n/2 \le R \le a_n$.

(e) There exist n_1, C_1, C_2 such that for $n \ge n_1$, $a_n/2 \le R \le a_n$ and $\epsilon \le |x| \le 1$,

$$\nu_{n,R}(x) \le C_1\,RQ'(R)/n \le C_2\,T(a_n)^{1/2}. \tag{4.38}$$

Proof

This is similar to that of Lemma 3.2 in [13], although there we set $R = a_n$. Note first that for $x \in [K, \infty)$, $(xQ'(x))' = Q'(x)T(x)$ is increasing, so $xQ'(x)$ is convex in $[K, \infty)$.

(a) The previous remark shows that for each fixed $x \geq K/R$,

$$\frac{RxQ'(Rx) - RsQ'(Rs)}{Rx - Rs}$$

is an increasing and positive function of $s \in [K/R,1]$.

(b) Since $(1 - x^2)^{1/2} \sim (1 - x)^{1/2}$ uniformly for $x \in (0,1)$, and since $s + x \sim 1$ uniformly for $\epsilon \leq x \leq 1$, we obtain (4.35) directly from (4.4) — note that $\nu_{n,R}$ is an even function.

(c),(d) For the given ϵ, choose η_1, η_2, and η_3 such that $0 < 1 - \epsilon < \eta_1 < \eta_2 < \eta_3 < 1$. Choose $\rho > 1$ such that $\rho(1 - \epsilon) < \eta_1$. For $x \in [0,1 - \epsilon]$ and $s \in [\eta_1,1]$, we have

$$RsQ'(Rs) \geq RsQ'(Rs) - RxQ'(Rx)$$

$$\geq RsQ'(Rs) \{1 - (x/s)^{T(Rx)}\} \quad \text{(by (3.2))}$$

$$\geq RsQ'(Rs) \{1 - ((1 - \epsilon)/\eta_1)^{T(0+)}\},$$

while

$$n \geq n(s^2 - x^2) \geq n\,(\eta_1^2 - (1 - \epsilon)^2).$$

We deduce that uniformly for $x \in [0,1 - \epsilon]$, $n \geq n_0$, and $a_n/2 \leq R \leq a_n$,

$$\frac{2}{\pi}\int_{\eta_1}^{1} \left\{\frac{1 - x^2}{1 - s^2}\right\}^{1/2} \frac{RsQ'(Rs) - RxQ'(Rx)}{n\,(s^2 - x^2)}\, ds$$

$$\sim n^{-1} \int_{\eta_1}^{1} RsQ'(Rs)\,(1 - s^2)^{-1/2}\, ds$$

$$\sim \frac{2}{n\pi}\int_0^1 RsQ'(Rs)\,(1 - s^2)^{-1/2}\, ds = 1 - B_{n,R}, \tag{4.39}$$

in view of (3.5), (4.5) and the monotonicity of $tQ'(t)$. Next, since the integrand in (4.39) (or (4.4)) is non–negative, we obtain

$$\nu_{n,R}(x) \geq C\,\{1 - B_{n,R}\}, \tag{4.40}$$

for $0 \leq x \leq 1 - \epsilon$, $n \geq n_0$, $a_n/2 \leq R \leq a_n$. In the other direction, for some C independent of n, x, R, let us write

$$\begin{aligned} J &:= \frac{2}{\pi}\int_0^{\eta_1} \{\frac{1 - x^2}{1 - s^2}\}^{1/2} \frac{RsQ'(Rs) - RxQ'(Rx)}{n\,(s^2 - x^2)}\,ds \\ &\leq C\,[\int_0^{x/\rho} + \int_{x/\rho}^{\rho x} + \int_{\rho x}^{\eta_1}]\, \frac{RsQ'(Rs) - RxQ'(Rx)}{n(s - x)(s + x)}\,ds \\ &=: I_1 + I_2 + I_3. \end{aligned} \tag{4.41}$$

Here, as $\rho > 1$,

$$\begin{aligned} I_1 &\leq C\,n^{-1} \int_0^{x/\rho} \frac{RxQ'(Rx)}{x(1 - 1/\rho)\,x}\,ds \\ &= C_2\,n^{-1}\,RQ'(Rx) \\ &\leq C_3\,n^{-1}\,RQ'(R\eta_1)\,(x/\eta_1)^{T(Rx)-1}, \end{aligned} \tag{4.42}$$

by (3.2). Next,

$$\begin{aligned} I_3 &\leq C_4 \int_{\rho x}^{\eta_1} RsQ'(Rs)/(ns^2)\,ds \\ &\leq C_4\,R\eta_1 Q'(R\eta_1) \int_{\rho x}^{\eta_1} (s/\eta_1)^{T(Rs)}/(ns^2)\,ds \quad (\text{by } (3.2)) \\ &\leq C_5\,R\eta_1 Q'(R\eta_1)\,n^{-1} \int_{x/\eta_1}^{1} t^{T(R\rho x)-2}\,dt, \end{aligned} \tag{4.43}$$

by the substitution $t = s/\eta_1$, and the monotonicity of T. Next,

$$\begin{aligned} I_2 &\leq C\,(\rho x - x/\rho)\,(nx)^{-1}\,R\max\{\tfrac{d}{du}(uQ'(u)) : u \in [Rx/\rho, Rx\rho]\,\} \\ &= C_6\,\frac{R}{n}\max\{\,T(u)Q'(u) : u \in [Rx/\rho, Rx\rho]\,\} \\ &\leq C_6\,\frac{R}{n}\,T(R\eta_1)\,Q'(R\eta_1)\,(x\rho/\eta_1)^{T(Rx/\rho) - 1}, \end{aligned} \tag{4.44}$$

by (3.2) and since $x\rho \leq (1 - \epsilon)\rho < \eta_1$. Here for $n \geq n_1$ and $a_n/2 \leq R \leq a_n$, the monotonicity of Q'' yields

$$T(R\eta_1)\,Q'(R\eta_1) = R\eta_1 Q''(R\eta_1) + Q'(R\eta_1)$$

$$\leq \eta_1(\eta_3 - \eta_2)^{-1} \int_{\eta_2}^{\eta_3} RQ''(Rt)\, dt + Q'(R\eta_1)$$

$$\leq C_7\, Q'(R\eta_3).$$

We deduce from (4.41) to (4.44) that

$$J \leq C_8\, R\eta_3 Q'(R\eta_3)\, n^{-1} \{(x\rho/\eta_1)^{T(Rx/\rho)-1} + \int_{x/\eta_1}^{1} t^{T(Rx/\rho)-2}\, dt\},$$

(4.45)

uniformly for $0 < x \leq 1 - \epsilon$, $n \geq n_1$, and $a_n/2 \leq R \leq a_n$. Next, from (3.5), we deduce that for $a_n/2 \leq R \leq a_n$,

$$\frac{1}{n} R\eta_3 Q'(R\eta_3) \leq C_9 \frac{2}{n\pi} \int_0^1 RsQ'(Rs)\, (1 - s^2)^{-1/2}\, ds = C_9\, \{1 - B_{n,R}\}.$$

From (4.39) and (4.45), and the fact that the term

$$(x\rho/\eta_1)^{T(Rx/\rho)-1}$$

in the right–hand side of (4.45) is easily seen to be no larger than a constant times the integral there, we deduce (4.37). If $x \geq \rho K/R$, so that $T(Rx/\rho) \geq 2$, then from (4.3), (4.39–41) and (4.45), we easily deduce (4.36).

If $T(0+) > 1$, then we see that (4.36) persists for $0 \leq x \leq K/R$.

(e) From (a) and (b) of this lemma, for $\epsilon \leq |x| \leq 1$, and n large enough,

$$\nu_{n,R}(x) \leq C \frac{R}{n} (1 - x)^{-1/2} \int_x^1 \psi_{n,R}(t)\, dt$$

$$\leq C \int_x^1 \frac{R}{n} \psi_{n,R}(t)\, (1 - t)^{-1/2}\, dt$$

$$\leq C_1 \int_x^1 \nu_{n,R}(t)\, (1 - t)^{-1}\, dt \leq C_1\, RQ'(R)/n,$$

by (b) of this lemma and (4.16). Finally from Lemma 3.2(c),

$$RQ'(R)/n \leq a_n Q'(a_n)/n \leq C_2 T(a_n)^{1/2}. \qquad \square$$

Concerning the behaviour of μ_{n,R_n} as $n \to \infty$, we prove the following, which (at least in spirit) is well known for the case $R_n = a_n$.

<u>Lemma 4.4</u>

Let $W \in SE$, except that we do not assume (2.3) or (2.4). Let $a_n/2 \le R_n \le a_n$, $n = 1, 2, 3, \dots$.

(a) Let $0 < \epsilon < 1/2$. Then, uniformly for $\epsilon \le |x| \le 1 - \epsilon$,

$$\lim_{n\to\infty} \mu_{n,R_n}(x) = \frac{1}{\pi}(1 - x^2)^{-1/2}. \tag{4.46}$$

(b) Uniformly in compact subsets of $\mathbb{C}\backslash[-1,1]$,

$$\lim_{n\to\infty} \int_{-1}^{1} \log|z - t|\, \mu_{n,R_n}(t)\, dt = \log|\varphi(z)/2|. \tag{4.47}$$

(c) Uniformly for $|z| \le 1$,

$$\lim_{n\to\infty} W(R_n|z|)^{1/n} = 1. \tag{4.48}$$

Furthermore,

$$\lim_{n\to\infty} \chi_{n,R_n}/n = \log 2. \tag{4.49}$$

(d) Uniformly in compact subsets of $\{z: |z| \le 1\}\backslash[-1,1]$,

$$\lim_{n\to\infty} U_{n,R_n}(z) = \log|\varphi(z)|. \tag{4.50}$$

<u>Proof</u>

(a) From (4.3), (4.4) and (4.5), and using a little manipulation, we obtain

$$\pi(1 - x^2)^{1/2}\mu_{n,R_n}(x) - 1$$

$$= \frac{2}{n\pi}\int_0^1 \frac{(1-s^2)R_n sQ'(R_n s) - (1-x^2)R_n xQ'(R_n x)}{s^2 - x^2}(1-s^2)^{-1/2}\, ds.$$

We can estimate this last right–hand side using the same methods as used for $\nu_{n,R}$ in the last lemma. For example, using the monotonicity of $tQ'(t)$, and $\epsilon \le |x| \le 1 - \epsilon$, we see that

$$\left|\frac{2}{n\pi}\left[\int_0^{\epsilon/2}+\int_{1-\epsilon/2}^{1}\right]\frac{(1-s^2)R_nsQ'(R_ns)-(1-x^2)R_nxQ'(R_nx)}{s^2-x^2}(1-s^2)^{-1/2}\,ds\right|$$

$$\leq C_1\,n^{-1}\left\{\int_{1-\epsilon/2}^{1}(1-s^2)^{1/2}\,R_nsQ'(R_ns)\,ds+R_nxQ'(R_nx)\right\}$$

$$\leq C_1\,n^{-1}\left\{\int_{1-\epsilon/2}^{1}R_nQ'(R_ns)\,ds+R_nxQ'(R_nx)\right\}$$

$$\leq C_2\,n^{-1}\{\,Q(R_n)+R_nQ'(R_n(1-\epsilon))\,\}.$$

Since $R_n \leq a_n$, Lemma 3.2 (a) and (b) show that this last right–hand side is o(1). The range $\epsilon/2 \leq |s| \leq 1-\epsilon/2$ may be dealt with as in the previous lemma – see (4.44).

(b) Now [19,p.217,Lemma 4.2],

$$\int_{-1}^{1}\log|z-t|\;\mu_{n,R_n}(t)\,dt-\log|\varphi(z)/2|$$

$$=\int_{-1}^{1}\log|z-t|\;\{\mu_{n,R_n}(t)-\tfrac{1}{\pi}(1-t^2)^{-1/2}\}\,dt.$$

Now (a) of this lemma shows that for each fixed $0<\epsilon<1/2$,

$$\lim_{n\to\infty}\int_{\epsilon\leq|t|\leq 1-\epsilon}\log|z-t|\;\{\mu_{n,R_n}(t)-\tfrac{1}{\pi}(1-t^2)^{-1/2}\}\,dt=0,$$

uniformly for z in a compact subset of $\mathbb{C}\backslash[-1,1]$. Next,

$$J_n(\epsilon):=\left\{\int_{|t|\leq\epsilon}+\int_{1-\epsilon\leq|t|\leq 1}\right\}\mu_{n,R_n}(t)\,dt$$

$$=1-\int_{\epsilon\leq|t|\leq 1-\epsilon}\mu_{n,R_n}(t)\,dt$$

$$\to 1-\int_{\epsilon\leq|t|\leq 1-\epsilon}\tfrac{1}{\pi}(1-t^2)^{-1/2}\,dt,$$

as $n\to\infty$. We deduce that for some C independent of ϵ, and $n \geq n_1(\epsilon)$,

$$J_n(\epsilon)\leq C\,\epsilon^{1/2},$$

and then for z in a compact subset of $\mathbb{C}\backslash[-1,1]$, we easily deduce a similar estimate for

$$\left[\int_{|t|\le\epsilon} + \int_{1-\epsilon\le|t|\le1}\right] \log|z-t| \; \{\mu_{n,R_n}(t) - \tfrac{1}{\pi}(1-t^2)^{-1/2}\}\, dt.$$

(c) Firstly, (4.48) follows from Lemma 3.2(a), since for $|z| \le 1$,

$$Q(R_n|z|)/n \le Q(a_n)/n \to 0 \text{ as } n \to \infty.$$

Recalling the definition (4.19) of $\chi_{n,R}$, we similarly obtain (4.49).

(d) This follows from (4.18), and from (b), (c) of this lemma. □

Next, we give a lemma on weighted polynomials that are approximately constant on $[-1,1]$.

<u>Lemma 4.5</u>

Let $W \in SE$, except that we do not assume (2.4). For $n \ge n_1$ and $a_n/2 \le R \le a_n$, there exist polynomials $P_n(x)$ of degree n, with leading coefficient $\exp(\chi_{n,R})$, with zeros $\{\hat{y}_{jn}\}_{j=1}^{n}$ satisfying

$$-1 =: \hat{y}_{0n} < \hat{y}_{1n} < \hat{y}_{2n} < \ldots < \hat{y}_{nn} < \hat{y}_{n+1,n} := 1,$$

and such that

$$C_1|x - \hat{y}_{cn}| \le |P_n(x)W(Rx)| \le C_2\, n^{C_3},\ x \in [-1,1], \tag{4.51}$$

where $\hat{y}_{cn} = \hat{y}_{cn}(n,R,x)$ denotes the closest zero of P_n to x. Further,

$$C_1 \le |P_n(x)W(Rx)| \exp(-nU_{n,R}(x)) \le C_2\, n^{C_3},\ |x| \in [1,2]. \tag{4.52}$$

The constants are independent of n, R and x. Concerning the spacing of $\{\hat{y}_{jn}\}_{j=1}^{n}$, there exists C_4 and C_5 s.t. for $n \ge n_1$ and $0 \le j \le n$,

$$\hat{y}_{j+1,n} - \hat{y}_{j,n} \ge n^{-C_5}. \tag{4.53}$$

<u>**Proof**</u>

Let

$$\Lambda(x) := \int_{-1}^{1} \log |x - t| \; \mu_{n,R}(t) \; dt, \; x \in \mathbb{C}.$$

First, (4.11) and Lemma 3.2(a) ensure that

$$|\Lambda(x)| \leq C, \tag{4.54}$$

uniformly for $x \in [-1,1]$, $n \geq n_1$, $0 < R \leq a_n$. Next, for $x > 1$,

$$\Lambda'(x) = \int_{-1}^{1} \mu_{n,R}(t) \; /(x - t) \; dt > 0,$$

so $\Lambda(x)$ is increasing in $(1, \infty)$. As for $|x| > 1$,

$$|\Lambda(x)| \leq \{|\log(|x| - 1)| + |\log(|x| + 1)|\} \int_{-1}^{1} \mu_{n,R}(t) \; dt$$

$$= |\log(|x| - 1)| + |\log(|x| + 1)|,$$

we obtain

$$|\Lambda(x)| \leq C, \; x \in [-2,2], \tag{4.55}$$

uniformly for $0 < R \leq a_n$, and $n \geq 1$, with C independent of n, R, x. Next, from (3.25), and from the bounds (4.9), (4.36), (4.37), and (4.38), we have uniformly for $n \geq n_1$, and $a_n/2 \leq R \leq a_n$,

$$\mu_{n,R}(x) \leq C_1 \int_{x}^{1} t^{T(0+)-2} \; dt, \; |x| \leq 1/2, \tag{4.56}$$

and

$$\mu_{n,R}(x) \leq C_1 \; \{T(a_n)^{1/2} + (1 - x^2)^{-1/2}\} \leq C_2 \; \{n^{1/29} + (1 - x^2)^{-1/2}\}, \tag{4.57}$$

$1/2 \leq |x| \leq 1$. It is then not difficult to see that if K is fixed, but large enough, we have

$$\int_{|x-t| \leq n^{-K}} |\log |x - t|| \; \mu_{n,R}(t) \; dt \leq (\log n)/(n + 1), \tag{4.58}$$

uniformly for $x \in [-1,1]$, $n \geq n_1$, and $a_n/2 \leq R \leq a_n$. We can then apply Lemma 6.1 in [14, p.40] to deduce that if the $\{\hat{y}_{jn}\}_{j=1}^{n}$ are chosen so that

$$\int_{\hat{y}_{jn}}^{\hat{y}_{j+1,n}} \mu_{n,R}(t)\,dt = \frac{1}{n+1}, \quad j = 0, 1, 2, \dots n, \tag{4.59}$$

and if

$$P_n^*(x) := \prod_{j=1}^{n} (x - \hat{y}_{jn}),$$

we have for $n \geq n_1$,

$$\tfrac{1}{4}|x - \hat{y}_{cn}| \leq |P_n^*(x)| \exp(-(n+1)\Lambda(x)) \leq n^{K+1}, \; x \in [-2,2], \tag{4.60}$$

and

$$|P_n^*(x)| \; \exp(-(n+1)\Lambda(x)) \geq \frac{1}{1+|x|}, \; |x| > 1. \tag{4.61}$$

Here, by (4.18), (4.19) and (4.20),

$$\exp(-n\Lambda(x)) = W(Rx) \exp(\chi_{n,R}), \; x \in [-1,1],$$

and

$$\exp(-n\Lambda(x)) = W(Rx) \exp(\chi_{n,R}) \exp(-nU_{n,R}(x)), \; |x| > 1.$$

Define

$$P_n(x) := \exp(\chi_{n,R}) \, P_n^*(x), \; n \geq n_1.$$

Then (4.55), (4.60) and (4.61) yield (4.51–2). Further, the definition (4.59) of the $\{\hat{y}_{jn}\}$, and (4.56–7) easily yield (4.53). □

We can rid ourselves of the troublesome factor $|x - \hat{y}_{cn}|$ in the lower bound in (4.51):

<u>Lemma 4.6</u>

Let $W \in SE$, except that we do not assume (2.4). Let $s \geq 0$, and $K > 0$, and

$$c_n = a_n(1 + s\,\Delta_n), \; n \geq 1. \tag{4.62}$$

Then there exist n_1 and $L > 0$, and $\hat{P}_n(x)$ of degree at most $n - K\,n\Delta_n$, $n \geq n_1$, such that

$$n^{-L} \le |\hat{P}_n(x)W(c_n x)| \le n^L, \; x \in [-1,1]. \tag{4.63}$$

<u>Proof</u>

First, let $P_n(x)$ be as in the previous lemma, with $R = a_n$. Further, let s_1 be a fixed, but large enough positive number. We shall need a lower bound for $U_{n,a_n}(x)$ for x close to 1. Since $B_{n,a_n} = 0$, (4.22) and the estimates of Lemma 4.2 show that for $n \ge n_1$, and $0 < \epsilon \le \epsilon_0$,

$$U_{n,a_n}(1+\epsilon)$$
$$\ge - C_1\{A_n^* \epsilon^{3/2} + T(a_n)^{3/2} \epsilon^{5/3} + T(a_n(1+\epsilon))^{3/2} (1+\epsilon)^{T(a_n(1+\epsilon))} \epsilon^2\}.$$

If $0 \le \epsilon \le s_1\Delta_n$, then by (3.34), for $n \ge n_1(s_1)$,

$$a_n(1+\epsilon) \le a_{2n},$$

so from (3.25), for $n \ge n_1$,

$$T(a_n)^{3/2} \le T(a_n(1+\epsilon))^{3/2} \le n^{3/29},$$

while

$$(1+\epsilon)^{T(a_n(1+\epsilon))} \le (1 + n^{-1/2})^{n^{2/29}} \le 2.$$

We deduce that for $0 \le \epsilon \le s_1 \Delta_n$, and $n \ge n_1$,

$$U_{n,a_n}(1+\epsilon) \ge - C_1\{A_n^* \epsilon^{3/2} + 3n^{3/29} \epsilon^{5/3}\}$$
$$\ge - C_1 A_n^* \epsilon^{3/2} \{1 + 3n^{3/29} A_n^{*-1} (s_1\Delta_n)^{1/6}\}$$
$$\ge - 2C_1 A_n^* \epsilon^{3/2},$$
$$\ge - 2C_1 \log n / n,$$

by (1.12). Then in addition to (4.51), we have from (4.52),

$$C_3 n^{-C_6} \le |P_n(x)W(a_n x)| \le C_4 n^{C_6}, \; 1 \le |x| \le 1 + s_1 \Delta_n, \tag{4.64}$$

for $n \ge n_1$. Now let $\alpha \ge C_5$, where C_5 is as in (4.53), and let

$$\mathcal{S}_n := (-1,1) \cap \bigcup_{j=1}^{n} (\hat{y}_{jn} - n^{-\alpha-2}, \hat{y}_{jn} + n^{-\alpha-2}).$$

From (4.51), we then have for $|x| \le 1 + s\Delta_n$ such that $x \notin \mathcal{S}_n$,

$$C_7 n^{-C_8} \le |P_n(x)W(a_n x)| \le C_4 n^{C_6}. \tag{4.65}$$

Now let $x \in \mathcal{S}_n$. We claim that then $x + n^{-\alpha-1}$ is not in $\mathcal{S}_n$. For, choosing j such that

$$|x - \hat{y}_{jn}| < n^{-\alpha-2},$$

we have from (4.53), for k = 1, 2, ... n,

$$|(x + n^{-\alpha-1}) - \hat{y}_{kn}| = |(x - \hat{y}_{jn}) + (\hat{y}_{jn} - \hat{y}_{kn}) + n^{-\alpha-1}|$$

$$\ge \begin{cases} n^{-\alpha} - n^{-\alpha-1} - n^{-\alpha-2}, & \text{if } k \ne j, \\ n^{-\alpha-1} - n^{-\alpha-2}, & \text{if } k = j, \end{cases}$$

$$\ge 2n^{-\alpha-2},$$

if also $n \ge 3$. Then from (4.65),

$$|P_n(x+n^{-\alpha-1})W(a_n x)| \ge C_7 n^{-C_8} W(a_n x)/W(a_n(x + n^{-\alpha-1})),\ x \in \mathcal{S}_n.$$

Here if $a_n x \ge r$, where $Q'(t)$ is increasing in $[r, \infty)$, then for some ξ between $a_n x$ and $a_n(x + n^{-\alpha-1})$,

$$1 \le W(a_n x)/W(a_n(x + n^{-\alpha-1})) = \exp(\ Q'(a_n\xi)\ a_n n^{-\alpha-1}\ )$$

$$\le \exp(\ Q'(a_{2n})\ a_{2n}\ n^{-\alpha-1}) = 1 + o(1),$$

if, as we can assume, $\alpha > 1$ – see Lemma 3.3(a) and (d). Similarly, since W is even and bounded above and below in each finite interval, we can show if $a_n|x| \le r$, that

$$1 \le W(a_n x)/W(a_n(x + n^{-\alpha-1})) \le C_0.$$

Defining

$$P_n^*(x) := P_n(x) + iP_n(x + n^{-\alpha-1}),$$

we have for $|x| \le 1 + s_1\Delta_n$, and $n \ge n_1$,

$$n^{-C_9} \le |P_n^*(x)W(a_n x)| \le n^{C_9}. \tag{4.66}$$

Now let K be a fixed, but large enough, positive number, and let us replace n in

(4.66), by $\ell := n - <2Kn\Delta_n>$, so that for n large enough,

$$\ell^{-C_9} \le |P^*_\ell(x)W(a_\ell x)| \le \ell^{C_9}, \; |x| \le 1 + s_1 \Delta_\ell. \tag{4.67}$$

Let us make the substitutions

$$a_\ell x = c_n u; \; \hat{P}_n(u) := P^*_\ell(x)$$

in (4.67). Note that $\hat{P}_n$ has degree at most $n - <2Kn\Delta_n>$. Since $\ell/n \to 1$ as $n \to \infty$, we obtain for some L, and $n \ge n_1$,

$$n^{-L} \le |\hat{P}_n(u)W(c_n u)| \le n^L, \; |u| \le (1 + s_1\Delta_\ell)a_\ell/c_n.$$

Here, by (3.22), and our choice (4.62) of c_n, for n large enough,

$$(1 + s_1\Delta_\ell)a_\ell/c_n \ge (1 + s_1\Delta_\ell)(1 - 2Kn\Delta_n)(1 - 2s\Delta_n) \ge 1,$$

if s_1 is large enough, since from Lemma 3.2(j),

$$\Delta_\ell \ge \left(\frac{\log \ell}{\ell A^*_n}\right)^{2/3} \ge \Delta_n/2,$$

n large enough. □

Finally, we recall a Markov–Bernstein Inequality:

Lemma 4.7

Let $W \in SE$, except that we do not assume (2.4). Then there exists C such that for $n \ge 1$ and $P \in \mathcal{P}_n$,

$$\|P'W\|_{L_\infty(\mathbb{R})} \le C\, Q'(a_n)\, \|PW\|_{L_\infty(\mathbb{R})}. \tag{4.68}$$

Proof

See [13, Theorem 1.3]. □

5 Infinite–Finite Range Inequalities

Inequalities relating L_p norms of weighted polynomials over finite and infinite intervals go back to Freud and Nevai (see [23]). An intensive investigation of the sharp form of these inequalities was initiated by Mhaskar and Saff [19,20]. See also [7,14,27]. The Erdös case, apart from general results in [20], has been considered in [7], but we need sharper results here. Theorems 5.1 and 5.2 are analogues of Theorems 7.1 and 7.2 in [14,p.49], and Theorem 5.1(a) appears in [14], while for the case of convex Q, it appeared in [20].

Theorem 5.1

Let $W \in SE$, except that we do not assume (2.4).

(a) Then for $n \geq 1$, and $P \in \mathcal{P}_n$,

$$\|PW\|_{L_\infty(\mathbb{R})} = \|PW\|_{L_\infty[-a_n,a_n]}, \tag{5.1}$$

and if P is not identically zero,

$$|PW|(x) < \|PW\|_{L_\infty(\mathbb{R})}, \ |x| > a_n. \tag{5.2}$$

Further for $0 < R \leq a_n$ and $z \in \mathbb{C}\backslash[-1,1]$,

$$|P(z)W(R|z|)| \leq \|P(x)W(Rx)\|_{L_\infty[-1,1]} \exp(nU_{n,R}(z)). \tag{5.3}$$

(b) Given $S > 0$, there exists $K > 0$ and $L > 0$, such that if

$$c_n := a_n(1 - K\Delta_n), \ n \geq 1, \tag{5.4}$$

then for n large enough,

$$n^L \geq \sup_{P \in \mathcal{P}_n} \{\|PW\|_{L_\infty(\mathbb{R})}/\|PW\|_{L_\infty[-c_n,c_n]}\} \geq n^S. \tag{5.5}$$

The proof is given below. The last part shows that a weighted polynomial $P(x)W(x)$, $P \in \mathcal{P}_n$, can grow by a power of n, in an interval of length $\sim a_n \Delta_n$ near a_n. For the L_p case, we can prove:

Theorem 5.2

Let $W \in SE$, except that we do not assume (2.4). Let $S > 0$, and $0 < p_0 < 1$. Then there exist K, L, $n_1 > 0$, such that if

$$c_n := a_n(1 + K\,\Delta_n),\ n \geq 1, \tag{5.6}$$

then uniformly for $n \geq n_1$ and $p_0 \leq p \leq \infty$,

$$n^{-S} \geq \sup_{P \in \mathcal{P}_n} \{\|PW\|_{L_p(|x|>c_n)}/\|PW\|_{L_p[-c_n,c_n]}\} \geq n^{-L}, \tag{5.7}$$

and

$$1 + n^{-S} \geq \sup_{P \in \mathcal{P}_n} \{\|PW\|_{L_p(\mathbb{R})}/\|PW\|_{L_p[-c_n,c_n]}\} \geq 1 + n^{-L}. \tag{5.8}$$

The proof of Theorem 5.2 is given below. Theorem 5.2 shows that after a distance of $a_n K\,\Delta_n$ beyond a_n, $(PW)(x)/\|PW\|_{L_p(\mathbb{R})}$ decays faster than any given negative power of n. A further distance of ϵa_n, $\epsilon > 0$ arbitrary, leads to geometric decay [14,19,20].

First,we need a lemma on the possible growth of $U_{n,R}$:

Lemma 5.3

Let $W \in SE$, except that we do not assume (2.4). Let $L > 1$. There exists n_1, C_1, C_2 such that if

$$R_n := a_n(1 - s\Delta_n),\ s \in [0,\log n],\ n \geq n_1, \tag{5.9}$$

then for $t \in [1,\log n]$ and $n \geq n_1$,

$$t^{1/2} (\log n) (C_1 s - t) \leq nU_{n,R_n}(1 + t\Delta_n)$$
$$\leq t^{1/2} (\log n) (C_2 s - t/2) + 1. \tag{5.10}$$

The upper bound in (5.10) holds also for $t \in [0,1]$, and the constants C_1, C_2 are independent of s, t and n.

<u>Proof</u>

This is similar to that of estimates in the proof of Theorem 7.1(iii) in [14,pp.51–53]. From (4.22),

$$nU_{n,R_n}(1 + t\Delta_n) = -n A_{n,R_n} \pi \sqrt{8} (t\Delta_n)^{3/2}/3 + n B_{n,R_n} \log \varphi(1 + t\Delta_n)$$
$$+ O(n \tau_{n,R_n} (t\Delta_n)^{5/3}) + O(n \rho_{n,R_n,t\Delta_n} (t\Delta_n)^2).$$

Recall that the constants in the order terms are independent of n, t, R_n and Q. Here from (4.27) and (5.9), and as $\log \varphi(1 + \epsilon) \sim \epsilon^{1/2}$, $\epsilon \to 0+$, we see that

$$n B_{n,R_n} \log \varphi(1 + t\Delta_n) \geq C_5 n s\Delta_n A^*_{n,R_n} (t\Delta_n)^{1/2},$$

where C_5 is an absolute constant. Next, from (4.28) and (3.25), for $t \in [1,\log n]$,

$$n \tau_{n,R_n} (t\Delta_n)^{5/3} = o(n^{1 + 3/29} (\log n)^{5/3} \{(\log n)/n\}^{10/9}) = o(1),$$

while from (4.29) and (3.34),

$$n \rho_{n,R_n,t\Delta_n} (t\Delta_n)^2 = O\{n\, T(a_{2n})^{3/2} (1 + (\log n) \Delta_n)^{T(a_{2n})} (\log n)^2 \Delta_n^2\}$$
$$= O\{n^{1 + 3/29} (1 + n^{-1/2})^{n^{1/2}} (\log n)^{2+4/3} n^{-4/3}\}$$
$$= o(1).$$

Then from these last estimates and (4.32), we obtain for $t \in [1,\log n]$ and $n \geq n_1$,

$$nU_{n,R_n}(1 + t\Delta_n) \geq n A^*_{n,R_n} \Delta_n^{3/2} t^{1/2} \{-0.95t + C_5 s\} + o(1)$$
$$= (\log n)\{1 - o(n^{3/29}(s\Delta_n)^{1/2})\}t^{1/2}\{-0.95t + C_5 s\} + o(1)$$
$$\text{(by (1.12), (4.30) and (3.25))}$$
$$\geq (\log n)\, t^{1/2} \{-t + C_5 s\},$$

if $n \geq n_1$, since also $t \geq 1$. Hence the lower bound in (5.10). The proof of the upper bound is very similar, one need only use the upper bound for $B_{n,R}$ in (4.27), rather than the lower bound. □

<u>Proof of Theorem 5.1</u>

(a) This was proved (in greater generality) in [14,Theorem 7.1,p.49].

(b) Let $S > 0$ be given. Choose fixed $s > 0$ and $t > 1$ such that

$$t^{1/2}\,(C_1 s - t) > 2S,$$

where C_1 is as in (5.10). Let R_n be given by (5.9). By Lemma 4.5, we can choose $P_n \in \mathcal{P}_n$ such that

$$|P_n(x)W(R_n x)| \le C_3\, n^{C_4},\ x \in [-1,1],$$

and

$$C_5 \le |P_n(x)W(R_n x)|\exp(-nU_{n,R_n}(x)) \le C_6 n^{C_7},\ |x| \in [1,2].$$

The constants are, in particular, independent of n and S. By Lemma 5.3, for $x_n := 1 + t\Delta_n$,

$$C_5\, n^{2S} \le |P_n(x_n)W(R_n x_n)| \le C_6 n^{C_8}.$$

Then, if

$$P_n^*(x) := P_n(x/R_n),$$

we have

$$\|P_n^* W\|_{L_\infty(\mathbb{R})} \,/\, \|P_n^* W\|_{L_\infty[-R_n,R_n]} \ge C_5 n^{2S}/(C_3 n^{C_4}) \ge n^S,$$

n large enough, provided S is large enough. This establishes the lower bound in (5.5), if we choose $K := s$ in (5.4), so that $c_n = R_n$. Next, for any $P \in \mathcal{P}_n$, (5.1) and (5.3) show that

$$\|PW\|_{L_\infty(\mathbb{R})} \,/\, \|PW\|_{L_\infty[-R_n,R_n]} = \|PW\|_{L_\infty[-a_n,a_n]} \,/\, \|PW\|_{L_\infty[-R_n,R_n]}$$
$$\le \max\{\ \exp(nU_{n,R_n}(x)) : x \in [1,a_n/R_n]\ \}$$

$$\leq n^L,$$

some L, by Lemma 5.3, and since $a_n/R_n \leq 1 + 2s\Delta_n$, n large enough. □

Lemma 5.4

Assume the notation and hypotheses of Theorem 5.2, and let $\{\ell_n\}_{n=1}^{\infty}$ be a sequence of positive integers satisfying

$$\lim_{n\to\infty} \ell_n^{27/29} /(n \log \ell_n) = \infty. \tag{5.11}$$

Then there exists n_1 such that uniformly for $p_0 \leq p \leq \infty$ and $n \geq n_1$,

$$I_{np} := \{\int_{a_{\ell_n}}^{\infty} \exp(npU_{n,a_n}(x/a_n))\, dx\}^{1/p} \leq \exp(-\tfrac{1}{4}\, \ell_n^{27/29}). \tag{5.12}$$

Proof

Note first that $\ell_n/n \to \infty$ as $n \to \infty$, so that $a_{\ell_n}/a_n \geq 1$ for n large enough. Now, from (4.7), (4.18), and (4.49), we see that for $n \geq n_1$, and $s \in [1, \infty)$,

$$U_{n,a_n}(s) \leq \log(s + 1) + 2\log 2 - Q(a_n s)/n =: f_n(s).$$

Here

$$f''_n(s) = -(s + 1)^{-2} - a_n^2 Q''(a_n s)/n < 0,\ s \in (1, \infty),$$

so $f'_n(s)$ is decreasing in $(1, \infty)$, while

$$f'_n(s) = (s + 1)^{-1} - a_n Q'(a_n s)/n \leq 1/2 - a_n Q'(a_n)/n \to -\infty,$$

as $n \to \infty$. Thus for $n \geq n_1$, $f'_n(s)$ is negative and decreasing in $(1, \infty)$, with limit $-\infty$ as $s \to \infty$. Then

$$I_{np} = a_n^{1/p} \{ \int_{a_{\ell_n}/a_n}^{\infty} \exp(npU_{n,a_n}(s))\, ds \}^{1/p}$$

$$\leq a_n^{1/p} \{ \int_{a_{\ell_n}/a_n}^{\infty} \exp(npf_n(s))\, ds \}^{1/p}$$

$$\leq a_n^{1/p} \{ \int_{a_{\ell_n}/a_n}^{\infty} \exp(npf_n(s)) \frac{npf_n'(s)}{npf_n'(a_n/a_{\ell_n})} ds \}^{1/p}$$

$$= a_n^{1/p} \{ \exp(npf_n(a_{\ell_n}/a_n)) / |npf_n'(a_{\ell_n}/a_n)| \}^{1/p}.$$

Next, from Lemma 3.2(a) (with j = 1), and from Lemma 3.2(g), for $n \geq n_1$,

$$f_n'(a_{\ell_n}/a_n) = (a_{\ell_n}/a_n + 1)^{-1} - a_n Q'(a_{\ell_n})/n \leq - \{a_n/a_{\ell_n}\}\{\ell_n/n\}$$

$$\leq - 1.$$

Next, from Lemma 3.3(e),

$$\exp(nf_n(a_{\ell_n}/a_n)) \leq \exp(n \log(a_{\ell_n}/a_n + 1) + 2n \log 2 - \ell_n^{27/29})$$

$$\leq \exp(- \tfrac{1}{2} \ell_n^{27/29}),$$

n large enough, by (5.11), and (3.19). Since $p^{-1/p}$ is bounded above for $p_0 \leq p \leq \infty$, and since

$$a_n^{1/p} \leq a_n^{1/p_0} \leq n^{1/p_0}, \; n \geq n_1,$$

we obtain the lemma. □

Proof of Theorem 5.2

With c_n given by (5.6), we obtain from (5.3), for $P \in \mathcal{P}_n$, and $p_0 \leq p < \infty$,

$$\|PW\|_{L_p(|x| \geq c_n)} \leq \|PW\|_{L_\infty(\mathbb{R})} \{2 \int_{c_n}^{\infty} \exp(npU_{n,a_n}(x/a_n))\, dx\}^{1/p}.$$

(5.13)

Now by a crude Nikolskii inequality [14,Lemma 7.3,p.53],

$$\|PW\|_{L_\infty(\mathbb{R})} \leq C\, n^{2/p_0} \|PW\|_{L_p(\mathbb{R})}, \tag{5.14}$$

where C is independent of n, $P \in \mathcal{P}_n$ and $p_0 \leq p \leq \infty$. Further, by (4.24) and by (5.10) (with s = 0, t = K), for $x \geq c_n$, $n \geq n_1$,

$$\exp(npU_{n,a_n}(x/a_n)) \leq \exp(npU_{n,a_n}(c_n/a_n))$$

$$\leq \exp(-p\, K^{3/2}/2 \log n + p)$$

$$\leq \exp(-p\, K^{3/2}/4 \log n), \tag{5.15}$$

if $n \geq n_1$, where n_1 is independent of $K \geq 1$, n and p. Now choose $\{\ell_n\}_{n=1}^{\infty}$ to be a sequence of positive integers such that

$$\lim_{n\to\infty} \ell_n^{27/29} / \{n (\log n)^2\} = 1.$$

Then we see that (5.11) is satisfied. From (5.15) and Lemma 5.4,

$$\{2 \int_{c_n}^{\infty} \exp(npU_{n,a_n}(x/a_n))\, dx\}^{1/p}$$

$$\leq \{2\, |a_{\ell_n} - c_n|\, \exp(-p\, K^{3/2}/4 \log n) + \exp(-\tfrac{p}{4}\, \ell_n^{27/29})\}^{1/p}$$

$$\leq \{2n^2 \exp(-p\, K^{3/2}/4 \log n) + \exp(-\tfrac{p}{8}\, n (\log n)^2)\}^{1/p},$$

if $n \geq n_1$, by Lemma 3.2 (g) and choice of $\{\ell_n\}_{n=1}^{\infty}$. From (5.13), (5.14), and this last estimate, we immediately deduce the upper bound in (5.7), uniformly for $p_0 \leq p < \infty$, and $n \geq n_1$, provided K is chosen large enough. The case $p = \infty$, follows by continuity and letting $p \to \infty$ (note that n_1 is independent of p). The upper bound in (5.8) then also follows, via the triangle inequality if $p \geq 1$. For $p \in [p_0, 1]$, a slightly more sophisticated argument is required. The lower bounds in (5.7–8) can be proved in a similar fashion to the lower bounds in Theorem 5.1. We spare the reader the details, noting that for Freud weights, a very similar proof appears in [14,Theorem 7.2,p.56]. □

6 Entire Functions

In [7], [14], a substantial role was played by the entire functions G_Q and $G_{Q/2}$ defined by (1.14) and (1.15). Here we need somewhat sharper asymptotics for G_Q, so must rework the proofs in [7,11]. This section is independent of the rest of this paper, and proceeds in greater generality, as we shall need these results in the appendix. In particular, the weights below include both Freud and Erdös type weights. We note that Theorem 6.1 generalizes and sharpens all the existing asymptotics for G_Q, notably those in [7,11].

<u>Theorem 6.1</u>

Let $W := e^{-Q}$, where Q is even, continuous in $\mathbb{R}$, $Q''(x)$ exists for large enough x, and $xQ'(x)$ is positive and increasing for large x, with limit ∞ as $x \to \infty$. Further assume that $T(x)$, defined by (1.13), satisfies

$$T(x) \geq C_1, \text{ x large enough,} \tag{6.1}$$

and

$$T(x) \sim xQ'(x)/Q(x), \text{ x large enough.} \tag{6.2}$$

Let j be a fixed non–negative integer.

(a) There exist C_1 and C_2 such that for x large enough,

$$C_1 \leq G_Q^{(j)}(x)\, W^2(x)\, (2\, Q'(x))^{-j}\, T(x)^{-1/2} \leq C_2. \tag{6.3}$$

(b) Suppose in addition that $Q'''(x)$ exists for x large enough, and that (2.5) holds for $k = 3$. Then as $x \to \infty$,

$$G_Q^{(j)}(x)\ W^2(x)\ (2Q'(x))^{-j}\ T(x)^{-1/2} = \pi^{1/2}\ \{1 + O(Q(x)^{-1/2})\}. \tag{6.4}$$

(c) Suppose in addition that $Q^{(4)}(x)$ exists for x large enough, and that (2.5) holds for $k = 3$ and 4. Then as $x \to \infty$,

$$G_Q^{(j)}(x)\ W^2(x)\ (2Q'(x))^{-j}\ T(x)^{-1/2} = \pi^{1/2}\ \{1 + O(Q(x)^{-1})\}. \tag{6.5}$$

We shall use Laplace's method and the Euler–Maclaurin summation formula. The "weak link" in our proofs that leads to the poor order terms in (6.4) and (6.5) is, surprisingly, Laplace's method for the dominant part of the series. First we need some preliminaries:

<u>Lemma 6.2</u>

Let W be as in Theorem 6.1(a). Then

(a) There exist C_1 and C_2 such that

$$Q'(x) \geq x^{C_1 - 1} \text{ and } Q(x) \geq x^{C_2},\ x \text{ large enough.} \tag{6.6}$$

(b) There exists C_1 such that

$$xQ'(x)/Q(x) \geq C_1,\ x \text{ large enough.} \tag{6.7}$$

(c) For u large enough,

$$q_u'/q_u = \{uT(q_u)\}^{-1}, \tag{6.8}$$

and

$$q_u Q'(q_u) = u. \tag{6.9}$$

(d) For u large enough,

$$T(q_u) \sim u/Q(q_u). \tag{6.10}$$

and

$$Q(q_u) \sim Q(q_{2u}). \tag{6.11}$$

(e) There exist C_3 and C_4 such that for u large enough,

$$C_3 \geq q_{2u}/q_u \geq 1 + C_4\, T(q_u)^{-1}. \tag{6.12}$$

(f) There exists C such that for x large enough,

$$T(x) \leq Q(x)^C. \tag{6.13}$$

<u>**Proof**</u>

(a) Since for x large enough,

$$Q''(x)/Q'(x) = (T(x) - 1)/x \geq (C_1 - 1)/x,$$

we obtain the inequality for Q' by integrating, and that for Q by integrating again.

(b) This follows directly from (6.1) and (6.2).

(c) Firstly (6.8) follows by differentiating (1.9), and using (1.13). Further, by definition of Freud's quantity $q_.$,

$$q_{uQ'(u)} Q'(q_{uQ'(u)}) = uQ'(u),$$

so the strict monotonicity of tQ'(t) yields (6.9).

(d) From (6.2),

$$T(q_u) \sim q_u Q'(q_u)/Q(q_u) = u/Q(q_u), \text{ u large enough.}$$

Next, from (6.8),

$$1 \leq Q(q_{2u})/Q(q_u) = \exp\left(\int_u^{2u} \frac{Q'(q_t)q_t'}{Q(q_t)}\, dt\right)$$

$$\leq \exp\left(\int_u^{2u} \frac{Q'(q_t)q_t}{Q(q_t)\; t\; T(q_t)}\, dt\right) \leq 2^{C_3},$$

from (6.2).

(e) $$q_{2u}/q_u = \exp\left(\int_u^{2u} q_t'/q_t\, dt\right) = \exp\left(\int_u^{2u} (tT(q_t))^{-1}\, dt\right)$$

$$\geq \exp\left(C_1 \int_u^{2u} Q(q_t)t^{-2}\, dt\right) \text{ (by (6.10))}$$

$$\geq \exp(C_1\, Q(q_u)\, (2u)^{-1})$$

$$\geq 1 + C_4/T(q_u),$$

by (6.10). Similarly using (6.1),

$$q_{2u}/q_u \le \exp(\int_u^{2u} (C_1 t)^{-1}\, dt) = 2^{1/C_1}.$$

(e) Now for x large enough,

$$\frac{d}{dx}(Q'/Q)(x) = Q''(x)/Q(x) - (Q'(x)/Q(x))^2$$

$$\le Q''(x)/Q(x) = \frac{Q'(x)}{xQ(x)}(T(x) - 1)$$

$$\le C\,(Q'(x)/Q(x))^2,$$

by (6.2). Dividing by $Q'(x)/Q(x)$, and then integrating, yields

$$\log(Q'(x)/Q(x)) \le C \log Q(x),$$

x large enough. Applying (6.2) again, and then (6.6), yields (6.13). □

Under more assumptions on Q, we can say more:

Lemma 6.3

Let W be as in Theorem 6.1(b).

(a) Then $T'(u)$ exists for u large enough, and

$$T'(u) = O(T^2(u)/u). \tag{6.14}$$

(b) There exist u_0 and C_1 s.t for $u \ge u_0$, $1/2 \le u/s \le 2$,

$$|T(q_u)/T(q_s) - 1| \le C_1 |u/s - 1|. \tag{6.15}$$

(c) For u large enough,

$$\frac{d}{du}\{q_u'/q_u\} = -\{u^2 T(q_u)\}^{-1}\{1 + q_u T'(q_u)/T^2(q_u)\} = O(1/\{u^2 T(q_u)\}). \tag{6.16}$$

Proof

(a) This is the same as in Lemma 3.4(a).

(b) Now by (6.8),

$$T(q_u)/T(q_s) = \exp\left(\int_s^u \frac{T'(q_t)}{T(q_t)} q_t'\, dt\right)$$

$$= \exp\left(\int_s^u \frac{T'(q_t)q_t}{T^2(q_t)t}\,dt\right)$$

$$= \exp\left(O\{|\log(u/s)|\}\right),$$

by (a) of this lemma. Then (6.15) follows.

(c) This follows from (6.8) by a straightforward calculation, and by using (a) of this lemma. □

<u>Lemma 6.4</u>

Let W be as in Theorem 6.1(c).

(a) Then for u large enough, T"(u) exists, and

$$T''(u) = O(T^3(u)/u^2). \tag{6.17}$$

(b) There exists u_0 and C_1 such that for $u \geq u_0$ and $1/2 \leq u/s \leq 2$,

$$|T'(q_u) - T'(q_s)| \leq C\, T^2(q_u)\, q_u^{-1}\, |u/s - 1|, \tag{6.18}$$

while

$$\left|\left(\frac{T'}{T}\right)(q_u) - \left(\frac{T'}{T}\right)(q_s)\right| \leq C\, T(q_u)\, q_u^{-1}\, |u/s - 1|. \tag{6.19}$$

(c) For u large enough,

$$\left(\frac{d}{du}\right)^2 (q_u'/q_u) = O(1/\{u^3\, T(q_u)\}). \tag{6.20}$$

<u>Proof</u>

(a) From (1.13), one obtains

$$T''(u) = 2Q'''(u)/Q'(u) - 2(Q''(u)/Q'(u))^2 + uQ^{(4)}(u)/Q'(u)$$
$$- 3uQ'''(u)Q''(u)/Q'(u)^2 + 2u(Q''(u)/Q'(u))^3.$$

Applying (6.1), (6.2) and (2.5) for k = 3, 4 leads to (6.17).

(b) This is a straightforward consequence of (6.17), (6.8), (6.12) and Lemma 6.3(b).

(c) By differentiating (6.16), one obtains

$$\left(\frac{d}{du}\right)^2 (q_u'/q_u)$$
$$= \{u^3 T(q_u)\}^{-1}\left[2 + 3\left\{\frac{T'(q_u)q_u}{T^2(q_u)}\right\}^2 + 3\frac{T'(q_u)q_u}{T^2(q_u)}\right.$$

$$- \frac{T'(q_u)q_u}{T^3(q_u)} - \frac{T''(q_u)q_u^2}{T^3(q_u)}] .$$

Applying (6.1), (6.2), (6.14) and (6.17), yields (6.20). □

Lemma 6.5

Let j and W be as in Theorem 6.1(a), and whenever defined, set

$$H(x,u) := j \log \left(\frac{2u}{x}\right) + \sum_{k=1}^{j-1} \log\left(1 - \frac{k}{2u}\right) + 2u \log\left(\frac{x}{q_u}\right) - \frac{1}{2} \log u + 2Q(q_u). \tag{6.21}$$

(a) Then, defining $\hat{J} := \max\{J, j/2\}$, we have for large enough x,

$$G_Q^{(j)}(x) = \sum_{n=\hat{J}}^{\infty} \exp(H(x,n)). \tag{6.22}$$

(b) If ', '' , and so on, denote differentiation with respect to u for fixed x, then

$$H'(x,u) = (j - 1/2)/u + \sum_{k=1}^{j-1} \frac{k}{2u^2} \frac{1}{1 - k/(2u)} + 2 \log\left(\frac{x}{q_u}\right), \tag{6.23}$$

where if j = 0 or j = 1, we interpret the sum as 0. Further,

$$H''(x,u) = -2 \frac{q_u'}{q_u} - (j - 1/2)/u^2 - u^{-3} P_2(1/u) \tag{6.24}$$

$$= - \frac{2}{uT(q_u)} \{1 + O(Q(q_u)^{-1})\}, \text{ u large enough,} \tag{6.25}$$

where $P_2(s)$ is a rational function of s, analytic in a neighbourhood of 0. Both H'' and P_2 are independent of x, and the constant in the order term in (6.25) is also independent of x.

(c) If W is as in Theorem 6.1(b), then

$$H'''(x,u) = O(1/\{u^2T(q_u)\}), \text{ u large enough;} \tag{6.26}$$

H''' is independent of x, as is the constant in the order term.

(d) If W is as in Theorem 6.1(c), then

$$H^{(4)}(x,u) = O(1/\{u^3T(q_u)\}),\ u \text{ large enough.} \tag{6.27}$$

$H^{(4)}$ is independent of x, as is the constant in the order term.

Proof

(a) This follows by differentiation of (1.14).

(b) Both (6.23) and (6.24) follow from differentiating (6.21), and using (1.9). Using (6.8) and (6.10), we obtain (6.25) from (6.24).

(c), (d) follow on differentiation of (6.24), and on applying (6.16), (6.20), and (6.10).

□

Lemma 6.6

Let j and W be as in Theorem 6.1(a), and assume the notation (6.21).

(a) There exist x_0 and $B > 0$ such that for $x \geq x_0$, there is a unique $y = y(x) \in [B, \infty)$ such that

$$H'(x,y) = 0. \tag{6.28}$$

Further,

$$x = q_y \exp\left(-\frac{(j-1/2)}{2y} - \sum_{k=1}^{j-1} \frac{k}{4y^2}\frac{1}{1-k/(2y)}\right). \tag{6.29}$$

(b) For $x \geq x_0$, we have the convergent expansion

$$(x - q_y)/x = \sum_{j=1}^{\infty} \hat{\alpha}_j\, y^{-j}, \tag{6.30}$$

where the $\hat{\alpha}_j$ are independent of x, y, and W. In particular,

$$\hat{\alpha}_1 = -(2j-1)/4\ ;\ \hat{\alpha}_2 = -(j^2 - j + 1/8)/4. \tag{6.31}$$

(c) We have, as $x \to \infty$,

$$y = xQ'(x)\ \{1 + O(Q(x)^{-1})\}; \tag{6.32}$$

(d) We have, as $x \to \infty$,

$$Q(q_y) = Q(x) - \hat{\alpha}_1 + O(Q(x)^{-1}); \tag{6.33}$$

and

$$H(x,y) = 2Q(x) - \frac{1}{2}\log(xQ'(x)) + j\log(2Q'(x)) + O(Q(x)^{-1}).$$

(6.34)

(e) For $x \geq x_0$,

$$-H''(x,y) \sim \{Q(x)T^2(x)\}^{-1}. \tag{6.35}$$

<u>Proof</u>

(a) From (6.25), there exists B independent of x, such that $H'(x,u)$ is decreasing for $u \in [B,\infty)$. (Recall that H'' is independent of x). Further, from (6.23), for each fixed $x > 0$,

$$\lim_{u\to\infty} H'(x,u) = -\infty,$$

while if x_0 is large enough, we see that $H'(x,B) > 0$ for $x \geq x_0$. Hence $y \in (B,\infty)$ exists and is unique. Then (6.29) follows from (6.23).

(b) This follows directly from (6.29).

(c) Now for some ξ between x and q_y, and from (6.30),

$$\begin{aligned}\log y &= \log q_y Q'(q_y)\\ &= \log\{xQ'(x)\} + (q_y - x)\frac{d}{dt}\log\{tQ'(t)\}_{|t=\xi}\\ &= \log\{xQ'(x)\} + O(x/y)\,T(\xi)/\xi.\end{aligned}$$

Here by (6.10) and (6.11), for $x \geq x_0$, and for $u \in [y/2,2y]$,

$$T(q_u) \sim u/Q(q_u)\begin{cases}\leq 2y/Q(q_{y/2})\\ \geq y/(2Q(q_y))\end{cases} \sim T(q_y).$$

Thus

$$T(q_u) \sim T(q_y),\ y \geq B,\ u \in [y/2,2y]. \tag{6.36}$$

Furthermore, (6.10) shows that

$$1/y = o(1/\min\{T(q_y),T(q_{y/2})\}),\ y \to \infty. \tag{6.37}$$

Then, from (6.30) and (6.12),

$$x = q_y(1 + O(1/y)) \in [q_{y/2},q_{2y}],\ x \geq x_0, \tag{6.38}$$

so

$$T(\xi) \sim T(q_y) \sim y/Q(q_y).$$

Thus

$$\log y = \log \{xQ'(x)\} + O(Q(q_y)^{-1}),$$

so

$$y = \{xQ'(x)\} \{1 + O(Q(q_y)^{-1})\}.$$

Then (6.32) follows, using (6.11) and (6.38).

(d) For some ξ between x and q_y,

$$\begin{aligned}
Q(q_y) &= Q(x) + (q_y - x)Q'(x) + (q_y - x)^2 Q''(\xi)/2 \\
&= Q(x) - xQ'(x) \{\hat{\alpha}_1/y\} \{1 + O(y^{-1})\} + O(y^{-2} x^2 T(\xi) Q'(\xi)/\xi) \\
&\qquad \text{(by (6.30) and (1.13), (6.1))} \\
&= Q(x) - \hat{\alpha}_1 + O(Q(x)^{-1} + y^{-1}) + O(y^{-2} x^2 T(\xi)^2 Q(\xi)/\xi^2) \\
&\qquad \text{(by (6.2) and (6.32))} \\
&= Q(x) - \hat{\alpha}_1 + O(Q(x)^{-1}),
\end{aligned}$$

for as in (c),

$$y^{-2} x^2 T(\xi)^2 Q(\xi)/\xi^2 = O(y^{-2} [y/Q(q_y)]^2 Q(q_y)) = O(Q(q_y)^{-1}),$$

since from (6.1) and (6.10),

$$Q(q_y) \sim y/T(q_y) = O(y).$$

Hence (6.33). Next, from (6.21), (6.29), (6.32) and (6.33),

$$\begin{aligned}
H(x,y) &= j \log(\tfrac{2y}{x}) + O(1/y) + 2y\log(x/q_y) - \tfrac{1}{2} \log y + 2Q(q_y) \\
&= j \log(2Q'(x)) - (j - 1/2) - \tfrac{1}{2} \log(xQ'(x)) + 2Q(x) - 2\hat{\alpha}_1 \\
&\quad + O(Q(x)^{-1} + y^{-1}).
\end{aligned}$$

Using (6.31), (6.1), and (6.2), we then obtain (6.34).

(e) From (6.25) and (6.10),

$$- H''(x,y) \sim \{y\, T(q_y)\}^{-1} \sim \{Q(q_y)\, T^2(q_y)\}^{-1}.$$

Then (6.33), (6.36) and (6.38) yield (6.35). □

Next, we express the dominant part of the series G_Q in terms of an integral, using the Euler–Maclaurin summation formula.

Lemma 6.7

Let j and W be as in Theorem 6.1(a), and assume the notation (6.21). Let K be some fixed, but large enough, positive number, and let

$$\omega := \omega(x) := \{K \log Q(q_y) / Q(q_y)\}^{1/2} y. \tag{6.39}$$

(a) Then

$$\omega \sim \{K Q(x) \log Q(x) T^2(x)\}^{1/2}, \tag{6.40}$$

where the constants in the $\sim$ are independent of x and K. Further, defining

$$\sum := \sum(n,x,\omega) := \sum_{|n-y| \le \omega} \exp(H(x,n)), \tag{6.41}$$

and

$$I := I(n,x,\omega) := \int_{y-\omega}^{y+\omega} \exp(H(x,t))\, dt, \tag{6.42}$$

we have as $x \to \infty$,

$$\sum = I \{1 + O(\frac{\log Q(x)}{Q(x)T^2(x)})\} + O(\exp[H(x,y)]\, Q(x)^{-100}). \tag{6.43}$$

(b) If in addition, W is as in Theorem 6.1(c), then as $x \to \infty$,

$$\sum = I \{1 + O[\frac{\log Q(x)}{Q(x)T^2(x)}]^2\} + O(\exp[H(x,y)]\, Q(x)^{-100}). \tag{6.44}$$

Proof

(a) Firstly, (6.40) follows from (6.32), (6.33) and (6.2). To prove (6.43), we use the Euler–Maclaurin summation formula [26,p.285]: If m is a positive integer, a, b are integers with $a < b$, and $f^{(2m)} \in L_1[a,b]$, then

$$\sum_{j=a}^{b} f(j) = \int_a^b f(t)\, dt + \tfrac{1}{2}\{f(a) + f(b)\} + \sum_{s=1}^{m} \frac{B_{2s}}{(2s)!} \{f^{(2s-1)}(b) - f^{(2s-1)}(a)\}$$
$$- \int_a^b \frac{B_{2m}(t - \langle t \rangle)}{(2m)!} f^{(2m)}(t)\, dt. \tag{6.45}$$

Here B_j and $B_j(t)$ are respectively Bernoulli numbers and Bernoulli polynomials, see [26,p.285]. We apply this with $m = 1$, $a := -\langle\omega\rangle$, $b := \langle\omega\rangle$ to

$$f(t) := \exp(H(x,t + \langle y\rangle)),\ t \in [a,b]. \tag{6.46}$$

Recall here that $\langle y\rangle$ denotes the greatest integer $\leq y$. Now

$$f'(t) = f(t)H'(x,t + \langle y\rangle),$$

$$f''(t) = f(t)\{H'(x,t + \langle y\rangle)^2 + H''(x,t+\langle y\rangle)\}.$$

Here, for some ξ between $t + \langle y\rangle$ and y, (6.25) shows that

$$H'(x,t+\langle y\rangle) = H'(x,y) + (t+ \langle y\rangle - y)\ H''(x,\xi)$$

$$= (t + \langle y\rangle - y)\left(\frac{-2}{\xi T(q_\xi)}\right)\{1 + O(Q(q_\xi)^{-1})\}.$$

Here, $t = O(\omega) = o(y)$, so $t + \langle y\rangle \sim y$. Then, (cf. (6.36) and (6.38))

$$T(q_\xi) \sim T(q_y) \sim T(x),$$

while from (6.2) and (6.32),

$$y \sim xQ'(x) \sim Q(x)T(x).$$

We deduce that

$$-H'(x,t + \langle y\rangle) \sim \frac{t + \langle y\rangle - y}{yT(x)} = O\left(\frac{\omega}{Q(x)T(x)^2}\right), \tag{6.47}$$

uniformly for $|t| \leq 2\omega$, $x \geq x_0(K)$. Similarly from (6.25),

$$-H''(x,t + \langle y\rangle) \sim \frac{1}{yT(x)} \sim \frac{1}{Q(x)T(x)^2}, \tag{6.48}$$

uniformly for $|t| \leq 2\omega$, $x \geq x_0(K)$. It is important to note here that the constants implicit in the $\sim$ are independent of K for x so large that $|\omega| \leq y/2$, that is for $x \geq x_0(K)$. We deduce that

$$|f''(t)| \leq C_1\ f(t)\ \frac{1}{Q(x)T(x)^2}\left\{\frac{\omega^2}{Q(x)T(x)^2} + 1\right\},$$

$$\leq C_2\ f(t)\ \frac{1}{Q(x)T(x)^2}\{K \log Q(x) + 1\}$$

uniformly for $|t| \leq 2\omega$, $K \geq K_0$, $x \geq x_0(K)$. Hence, uniformly for this range of parameters,

$$\left|\int_a^b \frac{B_{2m}(t-<t>)}{(2m)!} f^{(2m)}(t)dt\right| \le C_3 \frac{K \log Q(x)}{Q(x)\ T(x)^2} \int_a^b f(t)dt. \tag{6.49}$$

Next, we estimate f(a), f(b). Now for some ξ between $\pm <\omega> + <y>$ and y,

$$f(\pm<\omega>) = \exp[H(x,y) + (\pm <\omega> + <y> - y)^2 H''(x,\xi)/2]$$

$$\le \exp(H(x,y)) \exp\left(-C \frac{\omega^2}{Q(x)T(x)^2} \right) \text{ (by (6.35))}$$

$$\le \exp(H(x,y)) \exp(- C_1\ K \log Q(x)),$$

for $x \ge x_0(K)$, with C_1 independent of K. Similarly, we can use (6.47) to show that

$$|f'(\pm<\omega>)| \le C_2 \exp(H(x,y)) \exp(- C_1 K \log Q(x)) \left\{\frac{\log Q(x)}{Q(x)T(x)^2}\right\}^{1/2}.$$

If K is large enough, (6.45), (6.49) and these last estimates then yield

$$\sum_{j=a}^{b} \exp(H(x, j + <y>))$$

$$= \int_{-<\omega>}^{<\omega>} \exp(H(x,t+<y>))\ dt\ \left\{1 + O\left(\frac{\log Q(x)}{Q(x)T(x)^2}\right)\right\}$$

$$+ O(\exp(H(x,y)\ Q(x)^{-100}),$$

provided $x \ge x_0(K)$. Then (6.43) follows, noting that $\exp(H(x,t + <y>))$ is order $\exp(H(x,y))\ Q(x)^{-100}$, for $|t| \in [\omega-1,\omega+1]$, as is shown by our above estimates for $f(\pm<\omega>)$.

(b) This is similar to (a), so we merely outline the differences in the proof. We use (6.45) with m = 2 and the above choice of f. We see that

$$f'''(t) = f(t)\{H'(x,t+<y>)^3 + 3\ H'(x,t+<y>)H''(x,t+<y>)$$

$$+ H'''(x,t+<y>)\},$$

and

$$f^{(4)}(t) = f(t)\{H'(x,t+<y>)^4 + 6H'(x,t+<y>)^2 H''(x,t+<y>)$$

$$+ 4H'''(x,t+<y>)H'(x,t+<y>) + 3H''(x,t+<y>)^2 + H^{(4)}(x,t+<y>)\}.$$

Here, (6.47) and (6.48) remain valid, while, we can use (6.26) and (6.27) to deduce that

$$|H'''(x,t+\langle y\rangle)| = O(\frac{1}{Q(x)^2T(x)^3}) \; ; \; |H^{(4)}(x,t+\langle y\rangle)| = O(\frac{1}{Q(x)^3T(x)^4}),$$

and hence

$$|f^{(4)}(t)| \leq C\, f(t)\, \frac{(\log Q(x))^2}{Q(x)^2T(x)^4},$$

uniformly for $|t| \leq 2\,\omega$, and $x \geq x_0(K)$. The rest follows as before. □

Next, we deal with the tails of the series G_Q.

<u>Lemma 6.8</u>

Let j and W be as in Theorem 6.1(a), and assume the notation (6.21). Let K be some fixed, but large enough, positive number, and let ω be given by (6.39). Then for $x \geq x_0(K)$,

$$\sum^* := \sum^*(n,x,\omega) := \sum_{|n-y|>\omega} \exp(H(x,n)) = O(\exp[H(x,y)]\; Q(x)^{-100}). \tag{6.50}$$

<u>Proof</u>

Since for $x \geq x_0$, $H(x,u)$ is an increasing function of u for $u \in [B,y)$, and a decreasing function of u for $u \in (y,\infty)$, we can write

$$\sum^* \leq \int_{|u-y|\geq\omega-1} \exp(H(x,u))\,du + C_1(1+|x|)^{C_2},$$

where C_1 and C_2 are independent of n and x. The integral can be estimated exactly as was done in [7,Lemma 5.5, pp.49–50]. Further, from (6.34),

$$H(x,y) \sim Q(x),$$

it follows that as $x \to \infty$,

$$(1+|x|)^{C_2} = O(\exp(H(x,y))\; Q(x)^{-100}). \quad □$$

It is in the following lemma, in applying Laplace's method, that we obtain our worst order estimate:

<u>Lemma 6.9</u>

Assume the notation of Lemma 6.7.

(a) If W is as in Theorem 6.1(a), then there exist x_0, C_1 and C_2 such that for $x \geq x_0$,

$$C_1 \leq I \exp(-H(x,y)) \{yT(x)\}^{-1/2} \leq C_2. \tag{6.51}$$

(b) If W is as in Theorem 6.1(b), then as $x \to \infty$,

$$I \exp(-H(x,y)) (2\pi/|H''(x,y)|)^{-1/2} = 1 + O(Q(x)^{-1/2}). \tag{6.52}$$

(c) If W is as in Theorem 6.1(c), then as $x \to \infty$,

$$I \exp(-H(x,y)) (2\pi/|H''(x,y)|)^{-1/2} = 1 + O(Q(x)^{-1}). \tag{6.53}$$

<u>Proof</u>

(a) Recall that y is defined by (6.28). Then if $t \in (y-\omega,y+\omega)$, there exists $\xi = \xi(t,y)$ between t and y, s.t

$$H(x,t) = H(x,y) + H''(x,\xi)\,(t-y)^2/2.$$

Here, by (6.25) and (6.10) and by monotonicity of Q,

$$-H''(x,\xi) \sim Q(q_\xi)/\xi^2 \in [Q(q_{y-\omega})/(2y^2), 2Q(q_{y+\omega})/y^2].$$

Further, since $(y \pm \omega)/y \to 1$ as $y \to \infty$, (6.11) shows that

$$-H''(x,\xi) \sim Q(q_y)/y^2,$$

uniformly for $t \in (y-\omega,y+\omega)$ and $x \geq x_0$. Then

$$I = \int_{y-\omega}^{y+\omega} \exp(H(x,t))\,dt \leq \exp(H(x,y)) \int_{y-\omega}^{y+\omega} \exp\{-C_1\, Q(q_y)y^{-2}\,(t-y)^2\}\,dt$$

$$= \exp(H(x,y))\,\{Q(q_y)y^{-2}\}^{-1/2} \int_{-\ell}^{\ell} \exp(-\,C_1\,v^2)\,dv,$$

and similarly

$$I \geq \exp(H(x,y))\,\{Q(q_y)y^{-2}\}^{-1/2} \int_{-\ell}^{\ell} \exp(-\,C_2\,v^2)\,dv,$$

where C_1 and C_2 are independent of x and

$$\ell := \ell(x) := \{Q(q_y)y^{-2}\}^{1/2}\ \omega = \{K \log Q(q_y)\}^{1/2},$$

by (6.39). Since $\ell(x) \to \infty$ as $x \to \infty$, and since (6.32–33) show that

$$\{Q(q_y)y^{-2}\}^{-1/2} \sim \{Q(x)\ [xQ'(x)]^{-1}\ y^{-1}\}^{-1/2} \sim \{T(x)\ y\}^{1/2},$$

by (6.2), we obtain (6.51).

(b) If $t \in (y-\omega, y+\omega)$, then for some ξ between t and y,

$$H(x,t) = H(x,y) + H''(x,y)\ (t-y)^2/2 + H'''(x,\xi)\ (t-y)^3/3!.$$

Here, according to (6.26),

$$\begin{aligned}\beta &:= \max\{|H'''(x,u)| : |u-y| \le \omega\}/3! \le C_1 \max\{1/[u^2\ T(q_u)] : |u-y| \le \omega\}\\ &\le C_2\ y^{-2}\ T(q_y)^{-1} \quad \text{(by (6.36))}\\ &\le C_3\ Q(x)^{-2}\ T(x)^{-3},\end{aligned} \tag{6.54}$$

by (6.32), (6.33), (6.10) and (6.2). Then

$$\begin{aligned}I &\le \exp(H(x,y)) \int_{y-\omega}^{y+\omega} \exp[H''(x,y)\ (t-y)^2/2 + \beta\ |t-y|^3]\ dt\\ &= \exp(H(x,y))\ |2/H''(x,y)|^{1/2} \int_{-\ell}^{\ell} \exp(-v^2 + \rho\ |v|^3)\ dv,\end{aligned}$$

where by (6.35) and (6.40), and as above,

$$\begin{aligned}\ell := \ell(x) &:= |H''(x,y)/2|^{1/2}\ \omega\\ &\sim \{Q(x)\ T^2(x)\}^{-1/2}\ \{K \log Q(x)\ /Q(x)\ \}^{1/2}\ \{Q(x)\ T(x)\}\\ &\sim \{K \log Q(x)\}^{1/2},\end{aligned} \tag{6.55}$$

and (by (6.54))

$$\begin{aligned}\rho := \rho(x) &:= \beta\ |H''(x,y)/2|^{-3/2}\\ &= O(Q(x)^{-2}\ T(x)^{-3}\ \{Q(x)\ T^2(x)\}^{3/2}) = O(Q(x)^{-1/2}).\end{aligned}$$

Since $\rho\ell^3 = o(1)$, as $x \to \infty$, we can deduce that

$$\begin{aligned}I &\le \exp(H(x,y))\ |2/H''(x,y)|^{1/2}\{\int_{-\ell}^{\ell} \exp(-v^2)dv + 2\rho\int_{-\ell}^{\ell} \exp(-v^2)|v|^3\ dv\}\\ &= \exp(H(x,y))\ |2/H''(x,y)|^{1/2}\{\int_{-\infty}^{\infty} \exp(-v^2)dv + O(\exp(-\ell^2)/\ell) + O(\rho)\}\\ &= \exp(H(x,y))\ |2\pi/H''(x,y)|^{1/2}\{1 + O(Q(x)^{-1/2})\},\end{aligned}$$

if K is large enough. In exactly the same way, replacing ρ by $-\rho$, we obtain an analogous lower bound, and hence (6.52).

(c) Much as in (b),

$$I \leq \exp(H(x,y))\ |2/H''(x,y)|^{1/2} \int_{-\ell}^{\ell} \exp(-v^2 + \eta v^3 + \kappa v^4)\, dv,$$

where ℓ is given by, and satisfies, (6.55), where

$$\eta := H'''(x,y)\ |H''(x,y)/2|^{-3/2}/3!,$$

and

$$\kappa := \max\{|H^{(4)}(x,t)| : |t-y| \leq \omega\}\ |H''(x,y)/2|^{-2}/4!.$$

As before, we can use (6.25), (6.26) and (6.27) to deduce that

$$|\eta| \leq C_1\ Q(x)^{-1/2}\ ;\ \kappa \leq C_2\ Q(x)^{-1},$$

x large enough. As $\eta\ell^3$ and $\kappa\ell^4$ are o(1), we deduce that

$$I \leq \exp(H(x,y))\ |2/H''(x,y)|^{1/2} \{\int_{-\ell}^{\ell} \exp(-v^2)\, dv$$

$$+ \int_{-\ell}^{\ell} \exp(-v^2)\ \{\eta v^3 + \kappa v^4\}\, dv + 2\int_{-\ell}^{\ell} \exp(-v^2)\ \{\eta v^3 + \kappa v^4\}^2\, dv\}$$

$$= \exp(H(x,y))\ |2\pi/H''(x,y)|^{1/2} \{1 + O(Q(x)^{-1})\},$$

by our order estimates for η and κ, and since $\exp(-v^2)\, v^3$ is an odd function. As before, we can derive a similar lower bound. Hence (6.53). □

Proof of Theorem 6.1(a)

From Lemma 6.5(a), Lemma 6.7(a), and Lemma 6.8,

$$G_Q^{(j)}(x) = I\ \{1 + O(\frac{\log Q(x)}{Q(x)T^2(x)})\} + O(\exp(H(x,y))Q(x)^{-100}) \qquad (6.56)$$

$$\leq C_1 \exp(H(x,y))\ \{yT(x)\}^{1/2} + C_1\exp(H(x,y))Q(x)^{-100}$$

(by Lemma 6.9(a))

$$\leq C_2 \exp(2Q(x))\ (2Q'(x))^j\ \{yT(x)/[xQ'(x)]\}^{1/2}$$

(by (6.34))

$$\leq C_3 \exp(2Q(x))\ (2Q'(x))^j\ T(x)^{1/2},$$

by (6.32). In exactly the same way, we can derive a corresponding lower bound. Then (6.3) follows. □

Proof of Theorem 6.1(b)

By (6.56) and (6.52), as $x \to \infty$,

$$G_Q^{(j)}(x)$$
$$= \exp(H(x,y))\ (2\pi/|H''(x,y)|)^{1/2}\{1 + O(Q(x)^{-1/2})\}\{1 + O(\frac{\log Q(x)}{Q(x)T^2(x)})\}$$
$$+ O(\exp(H(x,y))Q(x)^{-100}). \tag{6.57}$$

We now obtain a more explicit expression for $H''(x,y)$. First, note that for some ξ between x and q_y,

$$T(q_y) = T(x) + (q_y - x)\ T'(\xi)$$
$$= T(x) + O[\ (x/y)\ T^2(\xi)/\xi\ ]\ \text{(by (6.30) and (6.14))}$$
$$= T(x) + O[\ T^2(x)/(xQ'(x))\ ]\ \text{(by (6.32), (6.36) and (6.38))}$$
$$= T(x)\{1 + O(1/Q(x))\}, \tag{6.58}$$

by (6.2). Next, from (6.25),

$$H''(x,y) = -\frac{2}{yT(q_y)}\{1 + O(Q(q_y)^{-1})\}$$
$$= -\frac{2}{xQ'(x)T(x)}\{1 + O(Q(x)^{-1})\}, \tag{6.59}$$

by (6.32), (6.33), and (6.58). Substituting into (6.57), and using (6.34) yields the desired result. □

Proof of Theorem 6.1(c)

From Lemma 6.5(a), Lemma 6.7(b), and Lemma 6.8,

$$G_Q^{(j)}(x) = I\ \{1 + O[\frac{\log Q(x)}{Q(x)T^2(x)}]^2\} + O(\exp(H(x,y))Q(x)^{-100})$$

$$= \exp(H(x,y))\ (2\pi/|H''(x,y)|)^{1/2}\{1 + O(Q(x)^{-1})\}\{1 + O[\frac{\log Q(x)}{Q(x)T^2(x)}]^2\}$$

$$+ O(\exp(H(x,y))Q(x)^{-100}),$$

by (6.53). We can now proceed exactly as in the proof of Theorem 6.1(b), applying the same estimates. □

7 Largest Zeros of Extremal Polynomials

In this section, we prove results for the largest zeros of extremal polynomials for Erdös weights and deduce Theorems 2.5 and 2.6. For Freud weights, analogues appeared in [14,Ch.8].

<u>Theorem 7.1</u>

Let $W \in SE$ and let j be a fixed positive integer.

(a) There exist n_1 and C such that uniformly for $2 \le p \le \infty$, and $n \ge n_1$,

$$|x_{jn}^{(p)}/a_n - 1| \le C\,\Delta_n. \tag{7.1}$$

(b) Let $0 < p_0 < 1$. There exist n_1 and C such that uniformly for $p_0 \le p \le \infty$, and $n \ge n_1$,

$$x_{1n}^{(p)}/a_n - 1 \le C\,\Delta_n. \tag{7.2}$$

Concerning the points of equioscillation $\{y_{jn}\}_{j=1}^{n+1}$ of $T_{n\infty}(W,x)W(x)$, we can prove:

<u>Corollary 7.2</u>

Let $W \in SE$ and let j be a fixed positive integer. There exists $C > 0$ such that

$$1 - C\Delta_n \le y_{jn}/a_n \le 1. \tag{7.3}$$

Corollary 7.3

Let $W \in SE$. Then

$$\sup_{P \in \mathcal{P}_n} \|P'W\|_{L_\infty(\mathbb{R})} / \|PW\|_{L_\infty(\mathbb{R})} \sim Q'(a_n) \sim (n/a_n)\, T(a_n)^{1/2}. \tag{7.4}$$

Proof of Theorem 7.1(b)

By Theorem 5.2, with c_n defined by (5.6), and K there large enough,

$$\|PW\|_{L_p(\mathbb{R})} \le (1+n^{-2})\, \|PW\|_{L_p[-c_n,c_n]},\ P \in \mathcal{P}_n,\ p_0 \le p \le \infty. \tag{7.5}$$

Let us suppose that for some $p_0 \le p \le \infty$, and large n, that

$$x_{1n}^{(p)} \ge a_n(1 + 3K\Delta_n),$$

and let

$$r(x) := (x - a_n\{1 + 2K\Delta_n\})/(x - x_{1n}^{(p)}).$$

It is not difficult to see that

$$\|r\|_{L_\infty[-c_n,c_n]} = |r(-c_n)| = (2 + 3K\Delta_n)/(1 + K\Delta_n + x_{1n}^{(p)}/a_n)$$

$$\le (2 + 3K\Delta_n)/(2 + 4K\Delta_n)$$

$$\le 1 - K\Delta_n/3,$$

n large enough. Then $T_{np}(W,x)r(x)$ is a monic polynomial of degree n, and by (7.5),

$$\|T_{np}(W,x)r(x)W(x)\|_{L_p(\mathbb{R})} \le (1 + n^{-2})\, \|T_{np}(W,x)r(x)W(x)\|_{L_p[-c_n,c_n]}$$

$$\le (1 + n^{-2})\,(1 - K\Delta_n/3)\, \|T_{np}(W,x)W(x)\|_{L_p[-c_n,c_n]}$$

$$< \|T_{np}(W,x)W(x)\|_{L_p(\mathbb{R})},$$

n large enough. Here, we have used Lemma 4.2(a) with $R = a_n$ and (3.25) to deduce that

$$A_n^* = A_{n,a_n}^* \le T(a_n) = o(n^{2/29}),$$

so

$$\Delta_n = \{(\log n)/(nA_n^*)\}^{2/3} \geq n^{-1}, \text{ n large enough.}$$

Thus we have a contradiction to extremality of $T_{np}(W,x)$. Hence, for $n \geq n_1$, and $p_0 \leq p \leq \infty$,

$$x_{1n}^{(p)} \leq a_n(1 + 3K\Delta_n). \qquad \square$$

In bounding certain quadrature sums, we shall need entire functions derived from those in Section 6.

Lemma 7.4

Let $W \in SE$, let $a > 0$, and let $b \in \mathbb{R}$. Then there exists an entire function

$$H(x) := 1 + \sum_{n=1}^{\infty} h_{2n}x^{2n},\ h_{2n} \geq 0,\ n \geq 1, \tag{7.6}$$

such that

$$H(x) \sim W(x)^{-a}\ T(x)^{1/2}\ Q(x)^{-b}, \text{ x large enough.} \tag{7.7}$$

Proof

We use Theorem 6.1(a), applied to

$$Q^*(x) := \tfrac{a}{2}\, Q(x) - \tfrac{b}{2} \log Q(x).$$

If necessary, modify $Q^*(x)$ for small $|x|$, so that Q^* is even and continuous in $\mathbb{R}$. Then for large enough x,

$$\begin{aligned} T^*(x) &:= \frac{(xQ^{*\prime}(x))'}{Q^{*\prime}(x)} \\ &= T(x) + \frac{b}{2}\left\{\frac{1}{Q(x)} + \frac{xQ'(x)}{Q^2(x)}\right\}/\left\{\frac{a}{2} - \frac{b}{2Q(x)}\right\} \\ &\sim T(x), \end{aligned}$$

by Lemma 3.1(i). Further, using (2.4), we see that

$$xQ^{*\prime}(x)/Q^*(x) \sim x\, Q'(x)/Q(x) \sim T(x) \sim T^*(x),$$

x large enough. Thus (6.1) and (6.2) hold for Q^*, and we can use $H := G_{Q^*}$, and

Theorem 6.1(a) with $j = 0$. □

Next, we show that a weighted polynomial is not too much larger than its values at the zeros of extremal polynomials.

Lemma 7.5

Let $W \in SE$. There exist n_1, C_3 and C_4 such that for $n \geq n_1$, $P \in \mathcal{P}_{n-1}$, $2 \leq p \leq \infty$, and $x \in \mathbb{R}$,

$$|P(x)W(x)| \leq C_3 n^{C_4} \max_{1 \leq j \leq n} |(PW)(x_{jn}^{(p)})/(x - x_{jn}^{(p)})| . \tag{7.8}$$

Proof

This follows Lemma 8.6 in [14,pp.62–64], and where appropriate, we refer to there for details. Let $2 \leq p < \infty$, $n \geq 1$, and

$$d\beta(x) = |p_{np}(W,x)|^{p-2} W^p(x)\, dx,\ x \in \mathbb{R}.$$

As in [14,Lemma 8.6], $p_{np}(W,x)$ is the orthonormal polynomial of degree n for $d\beta(x)$. Let $a := 2$, $b := 1/3$ in Lemma 7.4, and let H be the corresponding entire function. Proceeding exactly as in [14,Lemma 8.6,p.63], we deduce that for $x \in \mathbb{R}$,

$$|P(x)|$$

$$\leq C\, a_n \max_{1 \leq j \leq n} |(P/H^{1/2})(x_{jn}^{(p)}) p_{np}(W,x)/(x - x_{jn}^{(p)})|\ \{\int_{-\infty}^{\infty} H(x) d\beta(x)\}^{1/2}, \tag{7.9}$$

where C is independent of n, p, P and x. The non–negativity of the Maclaurin series coefficients of H is used in bounding a certain quadrature sum, see [14,p.63]. Now, by Hölder's inequality with parameters $2/p$ and $(p - 2)/p$, and by the normalisation (1.5),

$$\int_{-\infty}^{\infty} H(x) d\beta(x) = \int_{-\infty}^{\infty} H(x)\, W^2(x)\, |p_{np}(W,x)|^{p-2}\, W^{p-2}(x)\, dx$$

$$\leq \{\int_{-\infty}^{\infty} \{H(x)W^2(x)\}^{p/2}\, dx\}^{2/p}.$$

Next, from (7.7), for $|x|$ large enough,

$$H(x)\, W^2(x) \leq C_1 \{Q(x)^{-1/3}\, T(x)^{1/2}\}$$

$$\leq C_2 \{xQ'(x)\}^{-1/3}\, T(x)^{1/3+1/2} \qquad \text{(by (3.10))}$$

$$\leq C_3\, Q'(x)^{-1/3+1/15} \qquad \text{(by (2.3))}$$

$$\leq C_4\, x^{-4},$$

by Lemma 3.1(a). Thus, for some C_5 independent of p and x,

$$H(x)W^2(x) \leq C_5\, (1 + x^2)^{-2},\ x \in \mathbb{R},$$

so

$$\int_{-\infty}^{\infty} H(x)d\beta(x) \leq C_6 \{\int_{-\infty}^{\infty} (1 + x^2)^{-p}\, dx\}^{2/p} \leq C_7, \tag{7.10}$$

$2 \leq p \leq \infty$. Next, by a crude Nikolskii inequality [14,Lemma 7.3,p.53], we have that for some C independent of n, p, and W,

$$\|p_{np}(W,x)W(x)\|_{L_\infty(\mathbb{R})} \leq \{W(0)/W(2)\}\, (Cn^2)\, \|p_{np}(W,x)W(x)\|_{L_p(\mathbb{R})}$$

$$= C_8\, n^2. \tag{7.11}$$

Finally, from (7.7),

$$\max_{1\leq j\leq n}\ 1/\{H^{1/2}W\}(x_{jn}^{(p)})$$

$$\leq C_9\, [1 + \max\, \{Q(x_{jn}^{(p)})^{1/6}\, T(x_{jn}^{(p)})^{-1/4}\!: |x_{jn}^{(p)}| \geq C_{10}\}\,]$$

$$\leq C_{10}\, Q(a_n(1 + C\Delta_n))^{1/6} \qquad \text{(by Theorem 7.1(b))}$$

$$\leq C_{11}\, n^{1/6}, \tag{7.12}$$

by Lemma 3.2(a) with $j = 0$, and Lemma 3.3(d). Of course, the constants are independent of n, p and P. Combining (7.9) to (7.12), we obtain the lemma for $2 \leq p \leq \infty$. Finally let $p \to \infty$, and use the fact that the constants are independent of p, as well as the fact that

$$x_{jn}^{(p)} \to x_{jn}^{(\infty)},\ p \to \infty,\ j = 1,2,\ \dots\ n. \qquad \square$$

Proof of Theorem 7.1(a)

We proceed much as in [14], using the polynomials of Lemma 4.5 above, and the inequalities of Lemma 5.3 above. We shall prove by induction on j that

$$x_{jn}^{(p)}/a_n - 1 \geq - C\,\Delta_n, \tag{7.13}$$

$2 \leq p \leq \infty$, $n \geq n_1(j)$, for $j = 0, 1, 2, \ldots$. Together with Theorem 7.1(b), this will immediately yield the desired result. For $j = 0$, we set $x_{0n}^{(p)} = a_n$, and then (7.13) is immediate. Suppose now that we have proven (7.13) with $j - 1$ replacing j, for some positive integer j. Let us set

$$P(x) := S(x) \prod_{k=1}^{j-1} (x^2 - \{x_{kn}^{(p)}\}^2) \in \mathcal{P}_{n-2},$$

where S is an arbitrary element of $\mathcal{P}_{n-2j}$. If $j = 1$, the product is taken as 1. From Lemma 7.5, for some C_3 and C_4 independent of n, p, S, and for $|x| \leq x_{1n}^{(p)}$,

$$|S(x)W(x)| \leq C_3 n^{C_4} \{\prod_{k=1}^{j-1} |x^2 - \{x_{kn}^{(p)}\}^2|^{-1}\}\{2x_{1n}^{(p)}\}^{2j-2}$$

$$\times \max_{j\leq k\leq n+1-j} |(SW)(x_{kn}^{(p)})/(x-x_{kn}^{(p)})|\,.$$

Suppose that for some $K \geq 2$, $2 \leq p \leq \infty$, and given n,

$$x_{jn}^{(p)} \leq \min\,\{x_{j-1,n}^{(p)}, a_n\}\,(1 - 3K\,\Delta_n). \tag{7.14}$$

Let

$$I := (x_{j-1,n}^{(p)}(1 - 2K\Delta_n),\ x_{j-1,n}^{(p)}(1 - K\Delta_n)).$$

From (7.14) and (7.2), we obtain for some C_5 and C_6 independent of n, S, $K \geq 2$, and p,

$$|S(x)W(x)| \leq C_5\, n^{C_6} \max_{j\leq k\leq n+1-j} |(SW)(x_{kn}^{(p)})|\,,\ x \in I. \tag{7.15}$$

We now derive a contradiction, if K and n are too large. Let

$$R_{n-2j} := x_{j-1,n}^{(p)}(1 - 2K\,\Delta_n).$$

We claim that for $n \geq n_1(K)$, we may write

$$R_{n-2j} = a_{n-2j}(1 - \beta_n\,K\,\Delta_{n-2j}),\ \text{where } \beta_n \in [1/2,4]. \tag{7.16}$$

Firstly, from our induction hypothesis (7.13) for j − 1, and then from (3.22),

$$R_{n-2j}/a_{n-2j} \geq (a_n/a_{n-2j})\,(1 - C\,\Delta_n)\,(1 - 2K\,\Delta_n)$$
$$\geq (1 - 2j/n)\,(1 - C\,\Delta_n)\,(1 - 2K\,\Delta_n)$$
$$\geq 1 - 3K\Delta_n,$$

if K is large enough, and $n \geq n_1(K)$. In much the same way, we can use Theorem 7.1(b) to deduce that

$$R_{n-2j}/a_{n-2j} \leq 1 - K\Delta_n,$$

if K is large enough, and $n \geq n_1(K)$. Now exactly as in the proof of (3.46) of Lemma 3.4(d) – see (3.51) – we note that

$$A_n^* \geq A_{n-2j}^* \geq A_n^*\{1 - T(a_n)^{1/2}\,[1 - a_{n-2j}/a_n]^{1/2}\}$$
$$\geq A_n^*\{1 - C\,n^{1/29\,-\,1/2}\}.$$

Hence $A_n^*/A_{n-2j}^* \to 1$ as $n \to \infty$, and so $\Delta_n/\Delta_{n-2j} \to 1$ as $n \to \infty$. Then (7.16) follows. Next, by Lemma 4.5, there exists $S \in P_{n-2j}$ such that

$$|(SW)(R_{n-2j}u)| \leq C_7\,n^{C_8},\ u \in [-1,1], \tag{7.17}$$

and

$$|(SW)(R_{n-2j}u)| \geq C_9\,\exp((n-2j)U_{n-2j,R_{n-2j}}(u)),\ |u| \in [1,2]. \tag{7.18}$$

Of course, the constants are independent of n, p, K. Then (7.15) yields

$$C_9\,\exp((n-2j)\,U_{n-2j,R_{n-2j}}(u)) \leq C_5\,C_7\,n^{C_6+C_8}, \tag{7.19}$$

for $u > 1$ s.t $R_{n-2j}u \in I$. We claim that the set of such u contains an interval of the form $(1,1+\frac{1}{2}K\Delta_{n-2j}]$, n large enough. For

$$I = (R_{n-2j}, R_{n-2j}(1 - K\Delta_n)/(1 - 2K\Delta_n)),$$

and

$$(1 - K\Delta_n)/(1 - 2K\,\Delta_n) \geq 1 + \tfrac{3}{4}K\Delta_n \geq 1 + \tfrac{2}{3}K\Delta_{n-2j},$$

for $K \geq K_0$, and $n \geq n_1(K)$. Then, writing $u = 1 + t\Delta_{n-2j}$, $t \in (0,\frac{1}{2}K]$, Lemma 5.3

(with $s = \beta_n K$ and $n - 2j$ replacing n) yields

$$\exp((n-2j)\, U_{n-2j,R_{n-2j}}(u)) \geq \exp(t^{1/2} \log(n-2j)\, \{C_1 \beta_n K - t\})$$
$$\geq \exp(t^{1/2} \log(n-2j)\, \{C_1 K/2 - t\})$$

Choosing $t = K^{1/2}$, and K large enough, yields for $n \geq n_1(K)$, a contradiction to (7.19) since $C_5 \ldots C_9$ are independent of K. Hence for some $K > 0$, and for $2 \leq p \leq \infty$, and $n \geq n_1(K)$,

$$x_{jn}^{(p)} > \min\{x_{j-1,n}^{(p)}, a_n\}(1 - 3K\, \Delta_n).$$

Thus we have proven (7.13) for j. □

Proof of Corollary 7.2

The interlacing of the zeros and extrema of $T_{n\infty}(W,x)W(x)$, and Theorem 5.1(a), yield the result. □

Proof of Corollary 7.3

We have

$$\|T'_{n\infty}(W,x)W(x)\|_{L_\infty(\mathbb{R})} / \|T_{n\infty}(W,x)W(x)\|_{L_\infty(\mathbb{R})}$$
$$\geq |T'_{n\infty}(W,y_{1n})W(y_{1n})| / |T_{n\infty}(W,y_{1n})W(y_{1n})| = Q'(y_{1n}),$$

since

$$0 = \frac{d}{dx}\{T_{n\infty}(W,x)W(x)\}_{|\,x=y_{1n}}$$
$$= T'_{n\infty}(W,y_{1n})W(y_{1n}) - T_{n\infty}(W,y_{1n})W(y_{1n})Q'(y_{1n}).$$

Here, by Corollary 7.2, and then by Lemma 3.2(e), and (3.25),

$$a_n \geq y_{1n} \geq a_n\{1 - C\,\Delta_n\}$$
$$\geq a_{n(1 - n^{-1/2})}\{1 + C_1[\log(1 - n^{-1/2})^{-1}]\, /n^{2/29}\}\, \{1 - C\,\Delta_n\}$$
$$\geq a_{n(1 - n^{-1/2})},$$

n large enough. Hence, we can write $y_{1n} = a_{m(n)}$, where

$$n \geq m(n) \geq n(1 - n^{-1/2}).$$

Then (3.27) is satisfied, so by (3.28),

$$\lim_{n \to \infty} Q'(y_{1n})/Q'(a_n) = \lim_{n \to \infty} Q'(a_{m(n)})/Q'(a_n) = 1.$$

Hence,

$$\sup_{P \in \mathcal{P}_n} \|P'W\|_{L_\infty(\mathbb{R})}/\|PW\|_{L_\infty(\mathbb{R})} \geq CQ'(a_n).$$

The corresponding upper bound, appears in Theorem 1.3 of [13]. Finally, by Lemma 3.2(c),

$$Q'(a_n) \sim (n/a_n)\, T(a_n)^{1/2}. \quad \square$$

8 Approximation with Weighted Polynomials

Approximation by weighted polynomials of the form $P_n(x)W(a_nx)$, where W is a Freud weight, was intensively studied in [14]. For Erdös weights, the problem was considered in [7], but we need stronger results for this paper. The main result of this section is:

<u>Theorem 8.1</u>

Let $W \in SE$, and let $\hat{J}$ be a non–negative integer. Let

$$H(x) := 1 + \sum_{n=1}^{\infty} h_{2n} x^{2n}, \; h_{2n} \geq 0, \; n = 1, 2, 3, \dots , \tag{8.1}$$

be an entire function such that

$$H(x)\; W(x) \sim T(x)^{1/2}, \; x \text{ large}, \tag{8.2}$$

and for $j = 1, 2, \dots \hat{J}$,

$$| H^{(j)}(x)/H(x)| \leq C_1\; Q'(x)^j, \; x \text{ large}. \tag{8.3}$$

(If $\hat{J} = 0$, this assumption is omitted). For example, we may choose $H = G_{Q/2}$. Let h be bounded and analytic in $A := \{z: |z| < 1 \text{ and } |\mathrm{Im}\, z| < \delta\}$, some $\delta > 0$. Let $\eta \geq 0$, and let

$$g(x) := \begin{cases} (1 - x^2)^{\eta}\, h(x), & x \in (-1,1), \\ 0 & , x \in \mathbb{R}\backslash(-1,1). \end{cases} \tag{8.4}$$

Let y_{1n} be the largest point of equioscillation of $T_{n\infty}(W,x)W(x)$, and let

$$\xi_n := y_{1n}/a_n, \; n \text{ large enough}. \tag{8.5}$$

Then for $n \geq 2$, there exists $P_n \in \mathcal{P}_{n-1}$ such that for $x \in \mathbb{R}$,

$$|g(x) - P_n(x)/H(a_n x)|$$
$$\leq C_1 \|h\|_{L_\infty[-1,1]} \Delta_n^\eta \log\{1 + \frac{\Delta_n}{||x| - \xi_n|}\} T(a_n)^{1/2}, \tag{8.6}$$

and

$$\|g(x) - P_n(x)/H(a_n x)\|_{L_\infty(\mathbb{R})} \leq C_2 \|h\|_{L_\infty[-1,1]} \Delta_n^\eta (\log n)\ T(a_n)^{1/2}. \tag{8.7}$$

If $\hat{J} > 0$, then we have for $j = 1, 2, \ldots \hat{J}$, for $n \geq n_1$, and $|x| \leq \xi_n - 1/n$,

$$|(\tfrac{d}{dx})^j \{g(x) - P_n(x)/H(a_n x)\}|$$
$$\leq C_3 \|h\|_{L_\infty[-1,1]} \Delta_n^\eta \log\{1+\frac{\Delta_n}{||x|-\xi_n|}\} T(a_n)^{1/2} [nT(a_n)^{1/2}]^j. \tag{8.8}$$

The constants C_1, C_2 and C_3 are independent of n, x and h.

As in [14], we use Lagrange interpolation polynomials, and Hermite's contour integral error formula. Our choice of contour is simpler than that in Definition 11.3 of [14,p.82], and could have been used there as well.

<u>**Definition 8.2**</u>

With the notation of Theorem 8.1, let

$$\Gamma_{n1} := \{\xi_n e^{iy} : y \geq 0 \text{ and } 0 \leq \mathrm{Im}(\xi_n e^{iy}) \leq \epsilon\}, \tag{8.9}$$
$$\Gamma_{n2} := \{z: \mathrm{Im}\, z = \epsilon,\ |\mathrm{Re}\, z| \leq (\xi_n^2 - \epsilon^2)^{1/2}\}, \tag{8.10}$$

and let

$$\Gamma_n := \Gamma_{n1} \cup \overline{\Gamma_{n1}} \cup (-\Gamma_{n1}) \cup \overline{(-\Gamma_{n1})} \cup \Gamma_{n2} \cup \overline{\Gamma_{n2}}, \tag{8.11}$$

oriented in a positive sense.

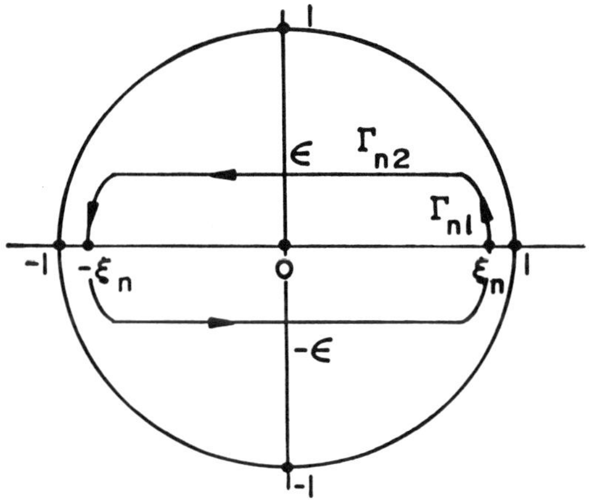

Figure 8.1: Γ_n

So Γ_n consists of two circular arcs, joined by horizontal line segments.

Lemma 8.3

Assume the notation and hypotheses of Theorem 8.1 and Definition 8.2. There exists n_1 such that if $n \geq n_1$ and $P_n(x)$ is the Lagrange interpolation polynomial of degree at most n–1 to $H(a_nx)g(x)$ at the zeros of $T_{n\infty}(W,a_nx)$, then

$$|g(x) - P_n(x)/H(a_nx)| \leq C_1 \|h\|_{L_\infty[-1,1]} I_n(|x|)\, T(a_n)^{1/2}, \; |x| < \xi_n, \tag{8.12}$$

and

$$|P_n(x)/H(a_nx)| \leq C_2 \|h\|_{L_\infty[-1,1]} I_n(|x|)\, T(a_n)^{1/2}, \; |x| > \xi_n. \tag{8.13}$$

Further, for j = 1, 2, ... $\hat{J}$, and for $|x| < \xi_n$,

$$\left|\left(\frac{d}{dx}\right)^j\{g(x) - P_n(x)/H(a_nx)\}\right| \le C_3 \|h\|_{L_\infty[-1,1]} I_n(|x|)\, T(a_n)^{1/2} [nT(a_n)^{1/2}]^j$$

$$\times \max\{1, (\xi_n - |x|)^{-1} n^{-1} T(a_n)^{-1/2}\}^j. \tag{8.14}$$

Here the constants are independent of n, x and h, and

$$I_n(x) := \int_{\Gamma_{n1}\cup\Gamma_{n2}} \frac{|1 - t^2|^\eta E_{n\infty}(W)}{|t - x|\,|T_{n\infty}(W, a_nt)W(a_n|t|)|} |dt|. \tag{8.15}$$

<u>Proof</u>

Inside and on Γ_n, we extend g to the complex plane by the first formula in (8.4). The Hermite error formula for Lagrange interpolation yields for $x \in (-\xi_n, \xi_n)$,

$$g(x) - P_n(x)/H(a_nx) = \frac{1}{2\pi i} \int_{\Gamma_n} \frac{g(t)T_{n\infty}(W,a_nx)/H(a_nx)}{(t-x)T_{n\infty}(W,a_nt)/H(a_nt)} dt, \tag{8.16}$$

while for $|x| > \xi_n$,

$$-P_n(x)/H(a_nx) = \frac{1}{2\pi i} \int_{\Gamma_n} \frac{g(t)T_{n\infty}(W,a_nx)/H(a_nx)}{(t-x)T_{n\infty}(W,a_nt)/H(a_nt)} dt. \tag{8.17}$$

Here for $t \in \Gamma_n$, $n \ge n_1$, and $x \in \mathbb{R}$, (8.2) yields

$$|H(a_nt)/H(a_nx)| \le H(a_n|t|)/H(a_nx)$$

$$\le C\,\{W(a_nx)/W(a_n|t|)\}\,\{T^{1/2}(a_n|t|)/T^{1/2}(a_nx)\}$$

$$\le C_1\,\{W(a_nx)/W(a_n|t|)\}\,T^{1/2}(a_n).$$

Here note that although we only assumed (8.2) for x large, it holds for all x, since both WH and T are bounded above and below by positive constants in each finite interval. Then for $|x| < \xi_n$,

$$|g(x) - P_n(x)/H(a_nx)|$$

$$\le C_2 T(a_n)^{1/2} \|h\|_{L_\infty[-1,1]} \int_{\Gamma_n} \frac{|1-t^2|^\eta E_{n\infty}(W)}{|t-x|\,|T_{n\infty}(W,a_nt)W(a_n|t|)|} |dt|. \tag{8.18}$$

Using the symmetry in the definition of Γ_n, we obtain (8.12), while (8.13) follows similarly. To prove (8.14), we differentiate (8.16):

$$(\frac{d}{dx})^j\{g(x) - P_n(x)/H(a_n x)\}$$

$$= \frac{1}{2\pi i}\int_{\Gamma_n} \frac{g(t)}{T_{n\infty}(W,a_n t)/H(a_n t)} (\frac{d}{dx})^j\{\frac{T_{n\infty}(W,a_n x)/H(a_n x)}{t - x}\} dt. \tag{8.19}$$

Now by the Markov–Bernstein inequality of Lemma 4.7,

$$\| T^{(k)}_{n\infty}(W,x)W(x)\|_{L_\infty(\mathbb{R})} \leq C\ Q'(a_n)^k \| T_{n\infty}(W,x)W(x)\|_{L_\infty(\mathbb{R})}$$

$$= C\ Q'(a_n)^k\ E_{n\infty}(W), \tag{8.20}$$

for $k = 1, 2, \ldots j$, and $n \geq n_1$. Next, it is easily seen by induction on m, that

$$(\frac{d}{dx})^m \{\frac{1}{H(a_n x)}\} = a_n^m \sum c\ H^{(\ell_1)}(a_n x)H^{(\ell_2)}(a_n x) \ldots H^{(\ell_m)}(a_n x)/H(a_n x)^{m+1},$$

where the coefficients c are independent of H and x, and the sum is over all m–tuples $(\ell_1,\ell_2, \ldots, \ell_m)$ of non–negative integers with $\ell_1 + \ell_2 + \ldots + \ell_m = m$. Then from (8.2) and (8.3), and the fact that T(x) is bounded below in $(0, \infty)$, we have for $|x| < \xi_n$,

$$|(\frac{d}{dx})^m \{\frac{1}{H(a_n x)}\}| \leq C_1\ \{a_n Q'(a_n)\}^m\ W(a_n x)$$

$$\leq C_2\ \{nT(a_n)^{1/2}\}^m\ W(a_n x), \tag{8.21}$$

by (3.14). Together with (8.20), this last estimate and Leibniz' formula yields for $|x| < \xi_n$,

$$|(\frac{d}{dx})^j\{\frac{T_{n\infty}(W,a_n x)/H(a_n x)}{t - x}\}|$$

$$= |\sum_{k=0}^{j} \sum_{m=0}^{j-k} \binom{j}{k}\binom{j-k}{m} (\frac{d}{dx})^k T_{n\infty}(W,a_n x)\ (\frac{d}{dx})^m\{1/H(a_n x)\}$$

$$\times\ (\frac{d}{dx})^{j-k-m}(t-x)^{-1}|$$

$$\leq E_{n\infty}(W)\ |t - x|^{-1}\ (nT(a_n)^{1/2})^j\ C_3\ \max\{1,\ (\xi_n-|x|)^{-1}\ n^{-1}\ T(a_n)^{-1/2}\}^j.$$

Then we obtain (8.14) from (8.19). □

Lemma 8.4

Assume the notation and hypotheses of Theorem 8.1.

(a) Uniformly in compact subsets of $\{z: |z| \le 1\}\backslash[-1,1]$, we have

$$\lim_{n\to\infty} a_n^{-1} |T_{n\infty}(W,a_n z)W(a_n|z|)|^{1/n} = |\varphi(z)|/2. \tag{8.22}$$

Further,

$$\lim_{n\to\infty} E_{n\infty}(W)^{1/n}/a_n = 1/2. \tag{8.23}$$

(b) If ℓ is a fixed positive integer, there exists n_1 such that for $n \ge n_1$, $z = \xi_n e^{iy}$, $0 \le y < \pi/2$, we have

$$E_{n\infty}(W)/|T_{n\infty}(W,a_n z)W(a_n|z|)| \le (1 + C_1 (y/\Delta_n)^2)^{-\ell}. \tag{8.24}$$

Proof

(a) This is standard, but we include the details. Let $P(x)$ be a monic polynomial of degree at most n. By (5.3),

$$|P(z)| W(a_n|z|)\exp(-nU_{n,a_n}(z)) \le \|P(x)W(a_n x)\|_{L_\infty[-1,1]}.$$

Letting $|z| \to \infty$, and taking account of the definition (4.18) of $U_{n,a_n}(z)$ yields

$$\exp(-\chi_{n,a_n}) \le \|P(x)W(a_n x)\|_{L_\infty[-1,1]}.$$

In the other direction, Lemma 4.5 constructs a polynomial $P_n(x)$ of degree n with leading coefficient $\exp(\chi_{n,a_n})$ satisfying (4.51) with $R = a_n$. Then $P_n^*(x) := \exp(-\chi_{n,a_n})P_n(x)$ is a monic polynomial satisfying

$$\|P_n^*(x)W(a_n x)\|_{L_\infty(\mathbb{R})} = \|P_n^*(x)W(a_n x)\|_{L_\infty[-1,1]} \le \exp(-\chi_{n,a_n})C_2 n^{C_3}.$$

In view of (4.49) in Lemma 4.4, we obtain

$$\lim_{n\to\infty} \{ \min_{\substack{P\in \mathcal{P}_n \\ P \text{ monic}}} \|P(x)W(a_n x)\|_{L_\infty(\mathbb{R})}\}^{1/n} = 1/2.$$

On making a substitution $a_n x = u$, we obtain (8.23). Next, define

$$f_n(z) := \frac{1}{n}\log|T_{n\infty}(W,a_nz)W(a_n|z|)| - U_{n,a_n}(z) - \frac{1}{n}\log E_{n\infty}(W).$$

Now $f_n(z)$ is harmonic in $\mathbb{C}\backslash[-1,1]$, (the $Q(a_n|z|)/n$ cancels) and by (5.3), $f_n(z) \leq 0$, $z \in \mathbb{C}\backslash[-1,1]$. Further, by (8.23), (4.18) and (4.49),

$$\lim_{n\to\infty} f_n(\infty) = \lim_{n\to\infty}\{\log a_n - \chi_{n,a_n}/n - \frac{1}{n}\log E_{n\infty}(W)\} = 0.$$

By a standard argument involving the maximum principle and normality, we have

$$\lim_{n\to\infty} f_n(z) = 0,$$

uniformly in compact subsets of $\mathbb{C}\backslash[-1,1]$. Then (8.23), (4.50), and the definition of $f_n(z)$ yield the result.

(b) We have for $z = \xi_n e^{iy}$, since $a_n\xi_n = y_{1n}$,

$$E_{n\infty}(W)/|T_{n\infty}(W,a_nz)W(a_n|z|)|$$

$$= |T_{n\infty}(W,a_n\xi_n)W(a_n\xi_n)|/|T_{n\infty}(W,a_nz)W(a_n\xi_n)|$$

$$= \prod_{x_{jn}^{(\infty)}>0}\left|\frac{(a_n\xi_n)^2 - (x_{jn}^{(\infty)})^2}{(a_nz)^2 - (x_{jn}^{(\infty)})^2}\right|$$

$$= \prod_{x_{jn}^{(\infty)}>0}\left|\frac{\xi_n^2 - z_{jn}^2}{z^2 - z_{jn}^2}\right|,$$

where $z_{jn} := x_{jn}^{(\infty)}/a_n$, $n \geq 1$. Here

$$|z^2 - z_{jn}^2|^2 = (\xi_n^2\cos 2y - z_{jn}^2)^2 + (\xi_n^2\sin 2y)^2$$

$$= (\xi_n^2 - z_{jn}^2)^2 + 2\xi_n^2 z_{jn}^2(1 - \cos 2y)$$

$$= (\xi_n^2 - z_{jn}^2)^2 + 4\xi_n^2 z_{jn}^2\sin^2 y.$$

In particular,

$$|z^2 - z_{jn}^2| \geq |\xi_n^2 - z_{jn}^2| = \xi_n^2 - z_{jn}^2.$$

Using the inequality

$$\sin y \geq \frac{2}{\pi}y,\ y \in [0,\pi/2],$$

we obtain for $n \geq n_1(\ell)$,

$$E_{n\infty}(W)/|T_{n\infty}(W,a_n z)W(a_n|z|)|$$

$$\leq \{\prod_{j=1}^{2\ell} \frac{(\xi_n^2 - z_{jn}^2)^2}{(\xi_n^2 - z_{jn}^2)^2 + 16\xi_n^2 z_{jn}^2 y^2/\pi^2}\}^{1/2},$$

$$= \prod_{j=1}^{2\ell} [1 + (\xi_n^2 - z_{jn}^2)^{-2} 16 \xi_n^2 z_{jn}^2 y^2/\pi^2]^{-1/2}$$

$$\leq [1 + C y^2/\Delta_n^2]^{-\ell},$$

for $n \geq n_1(\ell)$, by Theorem 7.1(a). □

Proof of Theorem 8.1

We shall estimate $I_n(x)$, given by (8.15), for $x \in [0, \infty)$. Firstly, uniformly for $x \in \mathbb{R}$,

$$I_{n2}(x)^{1/n} := \{\int_{\Gamma_{n2}} \left|\frac{1 - t^2}{t - |x|}\right|^{\eta} \frac{E_{n\infty}(W)}{|T_{n\infty}(W,a_n t)W(a_n|t|)|} |dt|\}^{1/n}$$

$$\leq \{\frac{C}{1 + |x|} \max_{t\in\Gamma_{n2}} \frac{E_{n\infty}(W)}{|T_{n\infty}(W,a_n t)W(a_n|t|)|}\}^{1/n}$$

$$\leq \{C/(1 + |x|)\}^{1/n} \{1/\min_{t\in\Gamma_{n2}} |\varphi(z)|^{-1} + o(1)\} < 1,$$

n large enough, by Lemma 8.4(a), and since Γ_{n2} is contained in the segment {z: Im z $= \epsilon$, $|\text{Re } z| \leq 1$}. Next, for $t \in \Gamma_{n1}$, $t = \xi_n e^{iy}$, we have if $x \in \mathbb{R}$, $n \geq n_1$,

$$|t - |x||^2 = (\xi_n \cos y - |x|)^2 + (\xi_n \sin y)^2$$

$$= (\xi_n - |x|)^2 + 2\xi_n |x| (1 - \cos y)$$

$$= (\xi_n - |x|)^2 + 4\xi_n |x| \sin^2(y/2)$$

$$\geq (\xi_n - |x|)^2 + 4/\pi^2 \xi_n |x| y^2,$$

and similarly

$$|1 - t^2|^2 = |1 - \xi_n^2 e^{2iy}|^2$$

$$= (1 - \xi_n^2)^2 + 4\,\xi_n^2 \sin^2 y \le C_1\,\Delta_n^2 + 4y^2,$$

by Corollary 7.2. Then using (8.9), Lemma 8.4(b), and choosing $\ell > \eta/2 + 1$,

$$I_{n1}(|x|) := \int_{\Gamma_{n1}} \left|\frac{1 - t^2}{t - |x|}\right|^{\eta} \frac{E_{n\infty}(W)}{|T_{n\infty}(W, a_n t) W(a_n |t|)|}\,|dt|$$

$$\le \int_0^{\arcsin(\epsilon/\xi_n)} \frac{(C_1\Delta_n^2 + 4y^2)^{\eta/2}}{\{(\xi_n - |x|)^2 + 4/\pi^2\,|x|\,y^2\}^{1/2}}\,\{1 + C_2(y/\Delta_n)^2\}^{-\ell} dy$$

$$\le C_3\,\Delta_n^{\eta} \int_0^{\arcsin(\epsilon/\xi_n)} \frac{\{1 + C_4(y/\Delta_n)^2\}^{-\ell+\eta/2}}{|\xi_n - |x|| + |x|^{1/2}\,y}\,dy$$

$$\le C_5\,\Delta_n^{\eta} \int_0^{\arcsin(\epsilon/\xi_n)\Delta_n^{-1}} \frac{(1 + C_4 u^2)^{-\ell+\eta/2}}{\Delta_n^{-1}\,|\xi_n - |x|| + |x|^{1/2} u}\,du.$$

Here, if $\Delta_n^{-1}\,|\xi_n - |x|| \ge 1$, we obtain

$$I_{n1}(|x|) \le C_6\,\Delta_n^{\eta+1} / |\xi_n - |x|| \int_0^{\infty} (1 + C_4 u^2)^{-\ell + \eta/2}\,du$$

$$\le C_7\,\Delta_n^{\eta} \log(1 + \Delta_n/|\xi_n - |x||).$$

If, however, $\Delta_n^{-1}\,|\xi_n - |x|| \le 1$, so that $|x|^{1/2} \ge 1/2$ for $n \ge n_1$, then

$$I_{n1}(|x|) \le C_8\{\Delta_n^{\eta} \int_0^1 1/\{\Delta_n^{-1}|\xi_n - |x|| + u\}\,du$$

$$+ \Delta_n^{\eta} \int_1^{\infty} (1 + C_4 u^2)^{-\ell+\eta/2}\,du\}$$

$$\le C_9\,\Delta_n^{\eta} \log\{1 + \Delta_n/|\xi_n - |x||\} + C_9\,\Delta_n^{\eta}$$

$$\le C_{10}\,\Delta_n^{\eta} \log\{1 + \Delta_n/|\xi_n - |x||\}.$$

Hence, for $x \in \mathbb{R}$, and $n \ge n_1$,

$$I_n(|x|) = I_{n1}(|x|) + I_{n2}(|x|)$$

$$\le C_{11}\,\Delta_n^{\eta} \log\{1 + \Delta_n/|\xi_n - |x||\}. \tag{8.25}$$

Then (8.6) follows from the estimates (8.12) and (8.13) of Lemma 8.3 for $|x| < \xi_n$ or $|x| > 1$. For $|x| \in [\xi_n, 1)$, we note that

$$|g(x)| \le C_{12} \|h\|_{L_\infty[-1,1]} \Delta_n^\eta,$$

while

$$\log\{1 + \Delta_n/|\xi_n - |x||\} \ge C_{13},$$

so (8.6) follows from (8.25), and (8.13) of Lemma 8.3. Further (8.8) follows from (8.14) for $|x| < \xi_n - 1/n$, noting that for such x,

$$(\xi_n - |x|)^{-1} n^{-1} T(a_n)^{-1/2} \le T(0)^{-1/2}.$$

Next, given arbitrarily large but fixed K, (8.6) yields

$$\|g(x) - P_n(x)/H(a_nx)\|_{L_\infty(\mathbb{R}\setminus(\xi_n - n^{-K}, \xi_n + n^{-K}))}$$

$$\le C_2(K) \|h\|_{L_\infty[-1,1]} \Delta_n^\eta \log n \, T(a_n)^{1/2}.$$

The range $(\xi_n - n^{-K}, \xi_n + n^{-K})$ can be filled exactly as was done in [14,pp.88–89]. One uses (5.3) and (5.10) to show that in an interval of length n^{-K} about ξ_n, (K fixed but large enough), $|P_n(x)W(a_nx)|$ cannot grow by a factor of more than 2, and hence $|P_n(x)/H(a_nx)|$ cannot grow by a factor of more than C $T(a_n)$. Then one uses a crude Markov–Bernstein inequality (Lemma 11.5 in [14,p.87]) to show that in an interval of length n^{-K}, (K fixed but large enough), $|P_n(x)/H(a_nx)|$ cannot change by more than $C\, a_n\, T(a_n)\, n^{2-K} = o(\Delta_n^\eta \log n)$. We omit the precise details, referring the reader to [14,pp.88–89]. Finally, note that $H = G_{Q/2}$ does indeed satisfy (8.1) to (8.3) by Theorem 6.1(a): The hypotheses (6.1) and (6.2) are respectively implied by (2.2) and (2.4). □

9 One-Sided Weighted Polynomial Approximation

In this section, we approximate $(1 - x^2)^{\kappa}$ from one side by weighted polynomials of the form $P_n(x)/H(c_n x)$, $x \in [-1,1]$. The main result of this section, Theorem 9.1, is rather unwieldy, but all the derivative and interpolatory conditions will be needed in establishing asymptotics for orthogonal polynomials and Christoffel functions on the real line.

<u>Theorem 9.1</u>

Let $W \in SE$. Let H be as in Theorem 8.1, with $\hat{J} = 2$. Fix $s \geq 0$, and let

$$c_n := a_n(1 + s\Delta_n), \ n \geq 1. \tag{9.1}$$

Further, let $\{k_n\}_{n=1}^{\infty}$ be a sequence of positive integers such that

$$k_n = O(n\Delta_n), \ n \to \infty. \tag{9.2}$$

Let $\kappa \in [-1,1]$. Then there exists $\delta > 0$ and n_1 such that for $n \geq n_1$ and $|x_0| \leq 1 - n^{-\delta}$, there exist polynomials $P_n(x)$ with real coefficients, of degree at most $2(n - k_n)$, positive in $[-1,1]$ except possibly for simple zeros at ± 1, and with the following properties: Let

$$\tau_n(u) := P_n(u)/H^2(c_n u)\ (1 - u^2)^{-\kappa}, \ u \in (-1,1). \tag{9.3}$$

Then we can ensure that either

$$\tau_n(u) \leq 1, \ u \in (-1,1) \text{ if } \kappa \in [-1,1), \tag{9.4}$$

or

$$\tau_n(u) \geq 1, \ u \in (-1,1), \text{ if } \kappa \in [0,1]. \tag{9.5}$$

Further,

$$\tau_n(x_0) = 1;\ \tau_n'(x_0) = 0; \tag{9.6}$$

and for j = 0, 1, 2,

$$|(\tfrac{d}{du})^j \{\tau_n(u) - 1\}| \le n^{-\delta},\ |u| \le 1 - n^{-\delta}. \tag{9.7}$$

Finally

$$\int_{-1}^{1} |\log \tau_n(t)|\ (1 - t^2)^{-1/2}\ dt \le n^{-\delta}, \tag{9.8}$$

and

$$(1 - u^2)\ n^{-1/\delta} \le \tau_n(u) \le (1 - u^2)^{-|\kappa|}\ n^{1/\delta},\ u \in [-1,1]. \tag{9.9}$$

For the P_n satisfying (9.5), or satisfying (9.4) with $\kappa < 0$, we may in addition ensure that P_n is positive in [−1,1].

Of course, δ above is independent of n, x_0, u. We can give a numerical value to δ, probably something like 10^{-4} would do, but this would complicate the proofs. Further, the bound (9.9) can be substantially improved, as the proofs indicate. One can also allow $\kappa = 1$ in (9.4), at the expense of lengthening the proofs.

We shall need several elementary approximation lemmas. The first constructs polynomials that are close to 1 in most of [−1,1] , but "zoom" up to m^K or down to m^{-K} near ± 1.

<u>Lemma 9.2</u>

Let $K \ge 1$, and let $\{\epsilon_m\}_{m=1}^{\infty} \subset (0,1/4)$. Let $\sigma := \pm 1$. There exists m_0 and for $m \ge m_0$ polynomials $R_m(x)$ of degree at most m, positive in [−1,1], such that

$$|R_m(x)/m^{\sigma K} - 1| \le C\, m^{-2},\ 1-\epsilon_m \le |x| \le 1; \tag{9.10}$$

$$|R_m(x) - 1| \le C\, m^{-2},\ |x| \le 1-\epsilon_m-m^{-1/4}, \tag{9.11}$$

and for j = 1, 2,

$$|R_m^{(j)}(x)| \le C\, m^{-1/5},\ |x| \le 1-\epsilon_m-m^{-1/4}. \qquad (9.12)$$

Finally,

$$1-Cm^{-2} \le \{R_m(x)\}^{\sigma} \le 2m^K,\ 1-\epsilon_m-m^{-1/4} \le |x| \le 1-\epsilon_m. \qquad (9.13)$$

Proof

Let $a_m := 1-\epsilon_m-m^{-1/4}$ and $b_m := 1-\epsilon_m$, and define an even function $f_m:[-1,1] \to [0,1]$ by

$$f_m(x) := \begin{cases} 0 & , \ 0 \le x \le a_m, \\ \left(\frac{x-a_m}{b_m-a_m}\right)^3 \{1+3\left(\frac{b_m-x}{b_m-a_m}\right)+6\left(\frac{b_m-x}{b_m-a_m}\right)^2\}, & a_m \le x \le b_m, \\ 1 & , \ b_m \le x \le 1. \end{cases}$$

It is easily seen that $f_m(x)$ is twice continuously differentiable in $[-1,1]$, with

$$\|f_m\|_{L_\infty[-1,1]} = 1,$$

$$\|f_m'\|_{L_\infty[-1,1]} \le 2/(b_m-a_m) = 2m^{1/4},$$

$$\|f_m''\|_{L_\infty[-1,1]} \le 15/(b_m-a_m)^2 = 15m^{1/2}.$$

Further, f_m''' is piecewise continuous with

$$\|f_m'''\|_{L_\infty[-1,1]} \le 90/(b_m-a_m)^3 = 90\, m^{3/4}.$$

From this last bound, we easily deduce estimates for the ordinary modulus of continuity of f_m'' on $[-1,1]$. By Jackson's Theorem, we can find polynomials $S_m(x)$ of degree at most $m/(64K \log m)$ such that for some constant $C = C(K)$, and for $j = 0, 1, 2,$

$$|f_m^{(j)}(x)-S_m^{(j)}(x)| \le C((\log m)/m)^{3-j}\, m^{3/4},\ x \in [-1,1]. \qquad (9.14).$$

Next, let $V_\ell(x)$ denote the $(\ell+1)$th partial sum of the Maclaurin series of e^x. By standard methods, one can show that for $j = 0, 1, 2$, and $\ell \ge \ell_0$,

$$|V_\ell^{(j)}(x)/e^x - 1| \le (1/2)^\ell,\ |x| \le \ell/16. \qquad (9.15)$$

Define

$$R_m(x) := V_\ell(\sigma K (\log m)\, S_m(x)),$$

where $\ell := 1 + <32K \log m>$. Then $R_m(x)$ has degree $\le \ell m/(64K \log m) \le m$, m large enough. Next, note that for $m \ge m_0$,

$$|\sigma K (\log m)\, S_m(x)| \le 2K \log m \le \ell/16, \quad x \in [-1,1].$$

Then (9.14) and (9.15) show that for $x \in [-1,1]$,

$$R_m(x)$$
$$= \exp(\sigma K(\log m)\,\{f_m(x) + O[(\log m)^3 m^{-9/4}]\})\,\{1 + O(1/2)^{32K\log m}\}$$
$$= \exp(\sigma K(\log m)\, f_m(x) + O(m^{-2}))\,(1 + O(m^{-2}))$$
$$= \begin{cases} m^{\sigma K}\{1 + O(m^{-2})\}, & 1 - \epsilon_m \le |x| \le 1, \\ 1 + O(m^{-2}) & , |x| \le 1 - \epsilon_m - m^{-1/4}, \end{cases}$$

by definition of f_m. Hence (9.10) and (9.11). Then (9.13) follows similarly using the fact that $0 \le f_m \le 1$ in $[-1,1]$. Next, for $|x| \le 1 - \epsilon_m - m^{-1/4}$,

$$R_m'(x) = V_\ell'(\sigma K \log m\, S_m(x))\, \sigma K \log m\, S_m'(x)$$
$$= \{1 + O(m^{-2})\}\, O[(\log m)^3 m^{-5/4}].$$

Similarly, we can estimate R_m''. □

Next, some specific one–sided approximations:

<u>Lemma 9.3</u>

Let $\tau \ge 0$, let $h(u) := (1 - u^2)^{-\tau}$, $u \in (-1,1)$, and let $S_m(u)$ be the (m+1)th partial sum of the Maclaurin series of h(u). There exists m_1, C_1 and C_2 such that for $m \ge m_1$,

$$|x_0| \le 1 - m^{-1/2}, \tag{9.16}$$

and

$$S_m^*(u) := S_m(u) + (h-S_m)(x_0) + (h-S_m)'(x_0)\,(u - x_0), \tag{9.17}$$

we have

$$S_m^{*(j)}(x_0) = h^{(j)}(x_0),\ j = 0,\ 1; \tag{9.18}$$

$$1/2 \le S_m^*(x) \le h(x),\ x \in (-1,1); \tag{9.19}$$

$$\|S_m^*\|_{L_\infty[-1,1]} \le C_1\ m^{\tau+1}; \tag{9.20}$$

and for j = 0, 1, 2,

$$|S_m^{*(j)}(x) - h^{(j)}(x)| \le \exp(-C_2 m^{1/2}),\ |x| \le 1 - m^{-1/2}. \tag{9.21}$$

<u>Proof</u>

If $\tau = 0$, $S_m^*(x) \equiv 1$, and the lemma follows immediately. Suppose now $\tau > 0$. Note that

$$S_m(x) = \sum_{j=0}^{m} \frac{\tau(\tau+1)...(\tau+j-1)}{j!}\ x^{2j},$$

so that

$$1 \le S_m(x) \le \sum_{j=0}^{\infty} \frac{\tau(\tau+1)...(\tau+j-1)}{j!}\ x^{2j} = h(x),\ x \in (-1,1). \tag{9.22}$$

Next, (9.18) follows immediately from the definition of $S_m^*(x)$. To prove the right inequality in (9.19), note that all Maclaurin series coefficients of h are non–negative, so the same is true of $h - S_m$. Then all even–order derivatives of $h - S_m$ are non–negative in (–1,1). Given $x \in (-1,1)$, the definition of S_m^* ensures that for some ξ between x and x_0,

$$(h - S_m^*)(x) = \tfrac{1}{2}\,(x - x_0)^2\ (h - S_m)''(\xi) \ge 0.$$

Hence the right inequality in (9.19).

Next, from the contour integral error formula for $h - S_m$, we have for $|x| \le 1 - m^{-1/2}$,

$$|h(x) - S_m(x)| = \Big|\frac{1}{2\pi i}\int_{|t|=1-1/m} \frac{h(t)}{t-x}\,\Big(\frac{x}{t}\Big)^{m+1} dt\Big|$$

$$\le C\, m^{\tau} \left(\frac{1 - m^{-1/2}}{1 - m^{-1}}\right)^{m+1} m^{1/2} \le e^{-C_2 m^{1/2}},$$

m large enough. Similarly, we can differentiate the contour integral with respect to x, to obtain

$$|h^{(j)}(x) - S_m^{(j)}(x)| \le e^{-C_2 m^{1/2}}, \; |x| \le 1 - m^{-1/2}, \; j = 1, 2.$$

Then (9.21) follows from (9.17). Further, the left inequality in (9.19) then follows from (9.22). Finally,

$$\frac{\tau(\tau+1)\ldots(\tau+j-1)}{j!} = \left|\binom{-\tau}{j}\right| = \left|\frac{1}{2\pi i}\int_{|t|=1-1/j} (1 - t)^{-\tau} t^{-j-1}\, dt\right|$$

$$\le C\, j^{\tau},$$

$j \ge 1$, whence

$$\|S_m\|_{L_\infty[-1,1]} \le C_3\, m^{\tau+1}.$$

Hence (9.20). □

Lemma 9.4

Let $0 \le \tau \le 1$, let $h(u) := (1 - u^2)^{\tau}$, $u \in (-1,1)$. Let $S_m(u)$ be the (m+1)th partial sum of h(u), $m \ge 1$. There exists m_1 and C_2 such that for $m \ge m_1$, x_0 satisfying (9.16), and $S_m^*(u)$ given by (9.17), we have (9.18) and (9.21). Further, for $m \ge m_1$, and $x \in (-1,1)$,

$$2 + \min\{\log m, \log (1 - x^2)^{-1}\} \ge S_m^*(x) \ge h(x). \tag{9.23}$$

Proof

If $\tau = 0$ or 1, then $S_m^*(u) \equiv S_m(u) \equiv h(u)$, and the result is immediate. Suppose now that $0 < \tau < 1$. Firstly, (9.18) is immediate, and (9.21) follows as in the previous lemma. Next, for $x \in (-1,1)$,

$$S_m(x) - h(x) = \sum_{j=m+1}^{\infty} \frac{\tau(1-\tau)(2-\tau)\ldots(j-1-\tau)}{j!} x^{2j},$$

and it follows that $S_m - h$ has all even order derivatives non–negative in (–1,1). Further, given $x \in (-1,1)$, there exists ξ between x and x_0 s.t

$$S^*_m(x) - h(x) = \frac{1}{2}(x - x_0)^2 (S_m - h)''(\xi) \geq 0.$$

Hence the right inequality in (9.23). Next,

$$0 \leq \frac{\tau(1-\tau)(2-\tau)\ldots(j-1-\tau)}{j!} \leq (j-1)!/j! = 1/j,$$

so for $|x| \leq 1$,

$$S_m(x) \leq 1 + \sum_{j=1}^{m} x^{2j}/j \leq \begin{cases} 2 + \log m, \\ 1 + \log(1 - x^2)^{-1}. \end{cases} \qquad \square$$

Next, we exhibit weighted polynomials that are not too small near ± 1.

Lemma 9.5

Assume the notation and hypotheses of Theorem 9.1. There exist L and n_1 such that for $n \geq n_1$, there exists a polynomial $\hat{P}_n$ of degree at most $n - k_n$ such that

$$n^{-L} \leq |\hat{P}_n(x)/H(c_n x)| \leq n^{-2}, \; x \in [-1,1], \tag{9.24}$$

and for j = 1, 2,

$$\left| \left(\frac{d}{dx}\right)^j \{\hat{P}_n(x)/H(c_n x)\} \right| \leq n^{-2}, \; x \in [-1,1]. \tag{9.25}$$

Proof

According to Lemma 4.6, (if necessary, multiplying the polynomials there by a suitable power of n) there exists L > 0 such that for n large enough, there exist polynomials $\hat{P}_n(x)$ of degree at most $n - k_n$, such that

$$n^{-L} \leq |\hat{P}_n(x)W(c_n x)| \leq n^{-8}, \; x \in [-1,1]. \tag{9.26}$$

Here since $c_n \geq a_n$,

$$\|\hat{P}_n(t/c_n)W(t)\|_{L_\infty(\mathbb{R})} \leq n^{-8},$$

so by Lemma 4.7, for j = 1, 2,

$$c_n^{-j} \|\hat{P}_n^{(j)}(t/c_n)W(t)\|_{L_\infty(\mathbb{R})} \leq C_1 n^{-8} Q'(a_n)^j.$$

Then using (3.24), we obtain

$$|\hat{P}_n^{(j)}(x)W(c_n x)| \le C_2\, n^{-5},\ x \in [-1,1],\ j = 0, 1, 2. \tag{9.27}$$

Next, exactly as at (8.21), we see that

$$|(\tfrac{d}{dx})^j\{1/H(c_n x)\}| \le C_3(n\, T(a_n)^{1/2})^j\, W(c_n x),\ x \in [-1,1], \tag{9.28}$$

$j = 0, 1, 2$. Combining (9.27), (9.28) with Leibniz' formula, yields (9.25). □

<u>Proof of Theorem 9.1</u>

This will consist of several steps, and after Step 5, we shall have to split the proof into different parts to obtain polynomials satisfying (9.4) or (9.5).

<u>Step1</u>: Approximate $(1 - x^2)^6$.

Let

$$g(x) := \begin{cases} (1 - x^2)^6, & x \in [-1,1], \\ 0 & \text{, otherwise.} \end{cases}$$

By (8.7) in Theorem 8.1, we can find polynomials $P_n(x)$ of degree at most n such that for $n \ge n_1$,

$$\|g(x) - P_n(x)/H(a_n x)\|_{L_\infty(\mathbb{R})} \le C_1 \{\frac{\log n}{n}\}^4 (\log n)\ T(a_n)^{1/2}$$

$$\le C_2\, n^{-3}. \tag{9.29}$$

Further, from (8.8) and Corollary 7.2, for $j = 1, 2$, and for $|x| \le 1 - C_3 \{(\log n)/n\}^{2/3}$,

$$|(\tfrac{d}{dx})^j\{g(x) - P_n(x)/H(a_n x)\}|$$

$$\le C_4 \{\frac{\log n}{n}\}^4\, T(a_n)^{1/2}\, (nT(a_n)^{1/2})^j \le n^{-3/2}. \tag{9.30}$$

Note that we chose $(1 - x^2)^6$ merely in order to ensure sufficiently fast rates of approximation, so as to control the size of the derivatives.

<u>Step 2</u>: From a_n to c_n and degree n to degree $n - <\alpha n\Delta_n>$.

Replace n in (9.29) and (9.30) by $\ell := n - <\alpha n\Delta_n>$, where α is a fixed but large enough positive number. We shall make there the substitutions

$$a_\ell x = c_n u \; ; \; P_\ell(x) =: P_n^*(u). \tag{9.31}$$

Note that $P_n^*(u)$ has degree at most $n - <\alpha n\Delta_n>$. From (9.29) (with n there replaced by ℓ),

$$|g(uc_n/a_\ell) - P_n^*(u)/H(c_n u)| \leq C_2 n^{-3}, \; u \in \mathbb{R}.$$

Here, for $u \in \mathbb{R}$,

$$|g(uc_n/a_\ell) - g(u)| \leq C_5 |c_n/a_\ell - 1|$$
$$\leq C_5 \{|\tfrac{\ell}{n} - 1| + s\Delta_n\} \leq C_6\{\tfrac{\log n}{n}\}^{2/3},$$

by (3.22), (9.1), and the choice of ℓ. Thus

$$\|g(u) - P_n^*(u)/H(c_n u)\|_{L_\infty(\mathbb{R})} \leq C_7\{\tfrac{\log n}{n}\}^{2/3}. \tag{9.32}$$

Since

$$(\tfrac{d}{du})^j = (\tfrac{a_\ell}{c_n})^j (\tfrac{d}{dx})^j,$$

and since $a_\ell/c_n \to 1$ as $n \to \infty$, similar reasonings and (9.30) yield

$$|(\tfrac{d}{du})^j\{g(u) - P_n^*(u)/H(c_n u)\}| \leq C_8 \{\tfrac{\log n}{n}\}^{2/3}, \; j = 1, 2, \tag{9.33}$$

for the range $|u| \leq (a_\ell/c_n)\{1 - C_3\{\frac{\log n}{n}\}^{2/3}\}$, and hence, as above, for the range

$$|u| \leq 1 - C_9 \{\tfrac{\log n}{n}\}^{2/3}. \tag{9.34}$$

<u>Step 3</u>: Approximate 1.

In the subsequent analysis, it will be easier to work with approximations to 1 rather than to $(1 - x^2)^6$. For this purpose, we use $S_m^*(u)$ of Lemma 9.3, with $\tau = 6$, $x_0 = 0$, and with $m := <n^{1/14}> + 1$. Let

$$R_n(u) := P_n^*(u)S_m^*(u),$$

a polynomial of degree at most $n - <\alpha n\Delta_n> + n^{1/14} + 1 \leq n - <\frac{\alpha}{2} n\Delta_n>$, n large enough. Next, from (9.21), (9.32) and (9.33), and some straightforward manipulations,

$$|(\frac{d}{du})^j \{R_n(u)/H(c_n u) - 1\}| \le C_{10}\, n^{-1/3}, \; j = 0, 1, 2, \tag{9.35}$$

$|u| \le 1 - n^{-1/30}$. Further, from (9.19), (9.20), and (9.32),

$$|R_n(u)/H(c_n u)| \le S_m^*(u)\{g(u) + C_{11}\, \{\frac{\log\, n}{n}\}^{2/3}\}$$
$$\le 1 + C_{12}\, n^{7/14}\, \{\frac{\log\, n}{n}\}^{2/3} \le 2, \; u \in [-1,1]. \tag{9.36}$$

<u>Step 4</u>: Interpolation at x_0.

For a given large enough n, let $f(u) := R_n(u)/H(c_n u)$, $u \in [-1,1]$, and define a linear polynomial

$$F(u) := F(u;x_0) := 1/f(x_0) - (u - x_0)\, (f'/f^2)(x_0).$$

Then defining

$$R_n^*(u) := R_n(u)F(u),$$

a polynomial of degree at most $n - <\frac{\alpha}{2}\, n\Delta_n> + 1$, we have

$$R_n^*(x_0)/H(c_n x_0) = 1; \tag{9.37}$$

and

$$\frac{d}{dx}\, \{R_n^*(x)/H(c_n x)\}_{|\, x\, =\, x_0} = 0. \tag{9.38}$$

Further, from (9.35), we deduce that

$$|(\frac{d}{dx})^j\{R_n^*(x)/H(c_n x) - 1\}| \le C_{13}\, n^{-1/3}, \; j = 0, 1, 2, \tag{9.39}$$

$|x|\, ;\, |x_0| \le 1 - n^{-1/30}$. Finally, from (9.36),

$$|R_n^*(x)/H(c_n x)| \le 3, \; x \in [-1,1];\; |x_0| \le 1 - n^{-1/30}. \tag{9.40}$$

<u>Step 5</u>: Beef up the weighted polynomials near ± 1.

We use the polynomials $\hat{P}_n(x)$ of Lemma 9.5 of degree at most $n - <\frac{\alpha}{4} n\Delta_n>$, n large enough. Define

$$Q_n(x) := R_n^*(x)^2 + \hat{P}_n(x)^2\, (x - x_0)^2,$$

a polynomial of degree at most $2n - 2<\frac{\alpha}{4}n\Delta_n>$, n large enough. From (9.39) and (9.24), (9.25),

$$|(\frac{d}{dx})^j\{Q_n(x)/H^2(c_nx) - 1\}| \le C_{14}\, n^{-1/3},\ j = 0,\ 1,\ 2, \tag{9.41}$$

$|x_0|$, $|x| \le 1 - n^{-1/30}$. Further, from (9.37) and (9.38),

$$Q_n(x_0)/H^2(c_nx_0) = 1; \tag{9.42}$$

$$\frac{d}{dx}\{Q_n(x)/H^2(c_nx)\}|_{x=x_0} = 0. \tag{9.43}$$

Suppose now that $n \ge n_1$ and $|x_0| \le 1 - 2n^{-1/30}$. Then for $|x| \le 1 - n^{-1/30}$, (9.39) ensures that

$$Q_n(x)/H^2(c_nx) \ge 1 - C_{13}\, n^{-1/3} \ge 1/2,$$

while if $|x| \ge 1 - n^{-1/30}$, then (9.24) shows that

$$Q_n(x)/H^2(c_nx) \ge n^{-1/15}\, \hat{P}_n(x)^2/H^2(c_nx) \ge n^{-1/15-2L}.$$

Hence, setting $\hat{L} := 1/15 + 2L$, and using (9.40),

$$n^{-\hat{L}} \le Q_n(x)/H^2(c_nx) \le 10,\ x \in [-1,1],\ |x_0| \le 1 - 2n^{-1/30}. \tag{9.44}$$

<u>Step 6</u> Polynomials satisfying (9.4).

By expanding $Q_n(x)/H^2(c_nx)$ about $x = x_0$, and using (9.41) – (9.43), we obtain for some ξ between x and x_0,

$$Q_n(x)/H^2(c_nx) = 1 + \frac{1}{2}(x - x_0)^2 (\frac{d}{dx})^2\{Q_n(x)/H^2(c_nx)\}|_{x=\xi}$$

$$\le 1 + C_{14}\, n^{-1/3}\,(x - x_0)^2, \tag{9.45}$$

$|x| \le 1 - n^{-1/30}$, $|x_0| \le 1 - 2n^{-1/30}$. Let us suppose first $\kappa < 0$ and let $m := <n^{|\kappa|/60}>$, and choose $S^*_m(x)$ to be the polynomial of Lemma 9.3 with $h(u) := (1 - u^2)^\kappa$, so that there $\tau = -\kappa$. Define

$$P_n(x) := Q_n(x)\{S^*_m(x) - n^{-1/6}(x - x_0)^2\},$$

a polynomial of degree $\le 2n - 2<\frac{\alpha}{4}n\Delta_n> + n^{1/60} \le 2n - <\frac{\alpha}{8}n\Delta_n>$, n large enough.

If α is large enough, it follows from (9.2) that $P_n(x)$ has degree at most $2n - 2k_n$, as required. Note first from (9.42–43), and from (9.18), that

$$\tau_n(x) := \{P_n(x)/H^2(c_n x)\}(1 - x^2)^{-\kappa},$$

satisfies the interpolation conditions (9.6). Next, from (9.19), and (9.44), for n large enough, $P_n(x)$ is positive in $[-1,1]$. Next, from (9.41) and (9.21), we see that (9.7) holds uniformly for $|u| \le 1 - n^{-\delta}$, $|x_0| \le 1 - n^{-\delta}$, δ small enough. Further, from (9.44) and (9.19), we obtain something better than the upper and lower bounds in (9.9). Then (9.8) follows from (9.9) and (9.7), possibly with a smaller δ. It remains to establish (9.4), and this is the hard part. First, from (9.45) and (9.19),

$$\tau_n(x) = \{Q_n(x)/H^2(c_n x)\} \{S_m^*(x)(1 - x^2)^{-\kappa} - n^{-1/6}(x - x_0)^2 (1 - x^2)^{-\kappa}\}$$

$$\le \{1 + C_{15}\, n^{-1/3}(x - x_0)^2\}\{1 - n^{-1/6}(x - x_0)^2(1 - x^2)^{-\kappa}\}$$

$$= 1 + n^{-1/3}(x-x_0)^2[C_{15} - n^{1/6}(1-x^2)^{-\kappa} - C_{15}n^{-1/6}(x-x_0)^2(1-x^2)^{-\kappa}]$$

$$\le 1 + n^{-1/3}(x - x_0)^2[C_{15} - n^{1/6}(1 - x^2)^{-\kappa}] \le 1,$$

if $|x| \le 1 - n^{-1/7}$, and $n \ge n_1$ (recall $-1 < \kappa < 0$). We have also used the fact that for n large enough,

$$\{x: |x| \le 1 - n^{-1/7}\} \subset \{x: |x| \le 1 - n^{-1/30}\}.$$

Also for $1 \ge |x| \ge 1 - n^{-1/7}$, we have from (9.20) and (9.44),

$$\tau_n(x) \le 10\, C_{16}\, m^{1-\kappa}\, (1 - x^2)^{-\kappa}$$

$$\le 10\, C_{17}\, n^{(1-\kappa)|\kappa|/60}\, n^{\kappa/7} \le 1,$$

$n \ge n_1$. Hence (9.4).

In the case $1 > \kappa \ge 0$, we take the above polynomials for $(1 - x^2)^{\kappa-1}$, and then absorb the extra factor $1 - x^2$ in τ into the polyonmial P_n. It is not difficult to see that all of (9.4), (9.6–9) are preserved. For the lower bound in (9.9), one needs to use that the polynomials for $(1 - x^2)^{\kappa-1}$ admit a bound like (9.44), and to recall the lower bound for S_m^* in (9.19).

<u>Step 7</u>: Polynomials satisfying (9.5).

This is more complicated than Step 6, as the polynomials of Step 5 may be too small near ± 1. We start with the polynomials $Q_n(x)$ satisfying (9.41–44), and multiply by the polynomials $R_m(x)$ of Lemma 9.2, with $m := <n^{1/9}>$; $\sigma = 1$; K fixed but large enough, and $\epsilon_m = n^{-1/40}$. More precisely, set

$$U_n(x) := Q_n(x)\ \{R_m(x) + (1 - R_m(x_0)) + \tfrac{d}{dt}(1 - R_m(t))|_{t=x_0}(x - x_0)\},$$

a polynomial of degree at most $2n - 2<\frac{\alpha}{8}\ n\Delta_n>$, n large enough. We have from (9.42) and (9.43),

$$\left.\begin{aligned} &U_n(x_0)/H^2(c_n x_0) = 1, \\ &(\tfrac{d}{dx})\{U_n(x)/H^2(c_n x)\}|_{x=x_0} = 0. \end{aligned}\right\} \tag{9.46}$$

Further, from (9.41), and (9.11), (9.12),

$$|(\tfrac{d}{dx})^j\{U_n(x)/H^2(c_n x) - 1\}| \le C_{18}\ n^{-1/50}, \tag{9.47}$$

for $j = 0, 1, 2, |x|, |x_0| \le 1 - 2n^{-1/40}$, since $m^{-1/4} \le n^{-1/40}$, n large enough. Next, from (9.10) and (9.44), if K in Lemma 9.2 is large enough,

$$U_n(x)/H^2(c_n x) \ge n^{-\hat{L}}(m^K - Cm^{-1/5}) \ge 1,\ 1 \ge |x| \ge 1-n^{-1/40},$$

provided $|x_0| \le 1 - 2n^{-1/40}$. Similarly, we obtain an upper bound. Hence,

$$1 \le U_n(x)/H^2(c_n x) \le n^S,\ 1 \ge |x| \ge 1-n^{-1/40},\ |x_0| \le 1-2n^{-1/40}. \tag{9.48}$$

Here $S > 0$ is independent of n, x, x_0. Further from (9.41), (9.11) and (9.13), we deduce that

$$1 - C_{19}\ n^{-2/9} \le U_n(x)/H^2(c_n x) \le n^S,\ 1-2n^{-1/40} \le |x| \le 1-n^{-1/40}. \tag{9.49}$$

Finally, we apply the polynomials S_m^* of Lemma 9.4, for $h(u) := (1-u^2)^\kappa$, with $m := \langle n^{1/4}\rangle$. More precisely, let $0 < \epsilon < 1$ be fixed, and let

$$P_n(x) := U_n(x)\{S_m^*(x) + n^{-\epsilon}(x-x_0)^2\},$$

a polynomial of degree at most $2n-2\langle\frac{\alpha}{16}n\Delta_n\rangle$, n large enough, which is positive in $[-1,1]$. For

$$\tau_n(x) := \{P_n(x)/H^2(c_n x)\}(1-x^2)^{-\kappa},$$

the interpolation conditions (9.6) follow from (9.18) and (9.46). Also from (9.21) and (9.47), we deduce that (9.7) holds for some $\delta > 0$. Next, (9.23), (9.47), (9.48) and (9.49) imply (9.9). Then (9.8) also follows.

It remains to prove (9.5). First, by (9.46), (9.47) and (9.23), for $|x| \le 1 - 2n^{-1/40}$ and $n \ge n_1$,

$$\tau_n(x) = \{U_n(x)/H^2(c_n x)\}\{S_m^*(x)(1-x^2)^{-\kappa} + n^{-\epsilon}(x-x_0)^2(1-x^2)^{-\kappa}\},$$

$$\ge \{1 - C_{18}n^{-1/50}(x-x_0)^2\}\{1 + n^{-\epsilon}(x-x_0)^2(1-x^2)^{-\kappa}\}$$

$$\ge 1 + n^{-\epsilon}(x-x_0)^2\{(1-x^2)^{-\kappa} - C_{18}n^{-1/50+\epsilon} - 4C_{18}n^{-1/50}(1-x^2)^{-\kappa}\}$$

$$\ge 1 + n^{-\epsilon}(x-x_0)^2\{\tfrac{1}{2}(1-x^2)^{-\kappa} - C_{18}n^{-1/50+\epsilon}\} \ge 1,$$

if $|x| \le 1 - 2n^{-1/40}$, and $\epsilon < 1/50$. Recall here that $\kappa \in [0,1]$. Thus, we have (9.5) for $|x| \le 1 - 2n^{-1/40}$. For $1 \ge |x| \ge 1 - 2n^{-1/40}$, we have from (9.48), (9.49), and (9.23), provided $|x_0| \le 1 - n^{-1/50}$,

$$\tau_n(x) \ge \{1 - C_{19}n^{-2/9}\}\{1 + n^{-\epsilon}(x-x_0)^2\,(1-x^2)^{-\kappa}\}$$

$$\ge \{1 - C_{19}n^{-2/9}\}\{1 + \tfrac{1}{2}\,n^{-\epsilon-1/25}\} \ge 1,$$

provided also $\epsilon < 2/9 - 1/25$. This completes the proof of (9.5). □

10 One-Sided Weighted Polynomial Approximation, II

In this section, we consider one–sided approximation of $(1-x^2)^{-\kappa}$ by $P_n(x)W(a_nx)$, (Theorem 10.1) and relative polynomial approximation of $1/\{W(a_nx)G_{Q/2}(a_nx)\}$ (Lemma 10.2).

<u>Theorem 10.1</u>

Let $W \in SE(3)$. Let $\{c_n\}_{n=1}^{\infty}$ and $\{k_n\}_{n=1}^{\infty}$ be as in Theorem 9.1. Let $\kappa \in (-1,1]$. For n large enough, there exist polynomials $S_n^+(t)$, $S_n^-(t)$ of degree at most $2(n-k_n)$, positive in $[-1,1]$ except possibly for simple zeros at ± 1, and with the following properties: Let

$$\tau_n^{\pm}(t) := W^2(c_nt)\, S_n^{\pm}(t)\, (1-t^2)^{-\kappa},\ t \in [-1,1]. \tag{10.1}$$

Then

$$\tau_n^+(t) \geq 1,\ t \in (-1,1),\ \text{if } \kappa \in [0,1], \tag{10.2}$$

and

$$\tau_n^-(t) \leq 1,\ t \in (-1,1),\ \text{if } \kappa \in (-1,1]. \tag{10.3}$$

Furthermore, there exists $\delta > 0$, such that for n large enough, and for $\tau_n := \tau_n^+$ or $\tau_n := \tau_n^-$,

$$|\tau_n(t) - 1| \leq n^{-\delta},\ |t| \leq 1 - n^{-\delta}, \tag{10.4}$$

$$\int_{-1}^{1} |\log \tau_n(t)|\ (1-t^2)^{-1/2}\, dt \leq n^{-\delta}, \tag{10.5}$$

and

$$n^{-1/\delta}(1-t^2) \leq \tau_n(t) \leq n^{1/\delta}\,(1-t^2)^{-|\kappa|},\ t \in (-1,1). \tag{10.6}$$

Further, given $\epsilon > 0$, we have for n large enough,

$$|\tau_n'(t)| \le n^{7/12},\ \epsilon/c_n \le |t| \le 1 - n^{-\delta}; \tag{10.7}$$

and

$$|\tau_n''(t)| \le n^{5/3},\ \epsilon/c_n \le |t| \le 1 - n^{-\delta}. \tag{10.8}$$

If Q' is continuous at 0, then (10.7) holds for $|t| \le 1 - n^{-\delta}$, while if Q" is continuous at 0, then (10.8) holds for $|t| \le 1 - n^{-\delta}$.

We remark that a numerical value may be given to δ, and that several of the statements may be strengthened, at the expense of further complicating the proofs. Of course, Theorem 10.1 is a corollary of Theorem 9.1, so we must pass from $1/H(c_n x)$ to $W(c_n x)$. This is achieved with the aid of the following lemma:

<u>Lemma 10.2</u>

Let $W \in SE(3)$, and let $\{c_n\}_{n=1}^{\infty}$ be as in Theorem 9.1. Let $G_{Q/2}(x)$ be defined by (1.15). Let

$$\Psi(x) := 1/\{G_{Q/2}(x)W(x)\},\ x \in \mathbb{R}. \tag{10.9}$$

Let m = m(n) satisfy for some C,

$$n \le m(n) \le Cn,\ n \text{ large enough}, \tag{10.10}$$

and let

$$\rho > 2/29. \tag{10.11}$$

Then there exists $\delta > 0$, and for n large enough, polynomials $R_n(x)$ of degree at most n^{ρ}, such that

$$|R_n(x)/\Psi(c_n x) - 1| \le n^{-\delta},\ |x| \le c_m/c_n; \tag{10.12}$$

and for each fixed $\epsilon > 0$,

$$\left|\frac{d}{dx}\{R_n(x)/\Psi(c_n x)\}\right| \le C_1\, n^{16/29}, \tag{10.13}$$

for $\epsilon/c_n \le |x| \le c_m/c_n$, and for the same range of x,

$$|(\tfrac{d}{dx})^2\{R_n(x)/\Psi(c_n x)\}| \le C_1\ n^{46/29}. \tag{10.14}$$

If Q' is continuous at 0, (10.13) is valid for $|x| \le c_m/c_n$, while if Q" is continuous at 0, (10.15) is valid for $|x| \le c_m/c_n$.

<u>Proof</u>

Note first that Theorem 6.1(b) is applicable, since (6.1) and (6.2) follow from (2.2) and (2.4). By Theorem 6.1(b),

$$\Psi(x) = \{\pi T(x)\}^{-1/2}\ \{1 + O(Q(x)^{-1/2})\},\ |x| \to \infty. \tag{10.15}$$

Further, it follows easily from (6.4) that

$$\Psi'(x) = O(Q'(x)/\{Q(x)T(x)\}^{1/2})$$
$$= O(\{Q'(x)/x\}^{1/2}),\ |x| \to \infty,$$

from (2.4) and (3.10). Also for x large enough,

$$\Psi''(x) = O(Q'(x)^2/\{Q(x)T(x)\}^{1/2} + Q''(x)/T(x)^{1/2}),$$
$$= O(Q'(x)^{3/2}/x^{1/2}),$$

since, by (3.10), and then by (6.2),

$$Q''(x)/T(x)^{1/2} \sim Q'(x)T(x)^{1/2}/x$$
$$\sim Q'(x)^{3/2}\ Q(x)^{-1/2}\ x^{-1/2}.$$

Hence,

$$|\Psi'(x)| \le C_1\ Q'(x)^{1/2}/x^{1/2},\ x \ge C_2, \tag{10.16}$$

and

$$|\Psi''(x)| \le C_1\ Q'(x)^{3/2}/x^{1/2},\ x \ge C_2. \tag{10.17}$$

Let us write

$$\Psi(x) =: Z(x)\ \{\pi T(x)\}^{-1/2}, \tag{10.18}$$

where according to (10.15),

$$Z(x) = 1 + O(Q(x)^{-1/2}),\ |x| \to \infty. \tag{10.19}$$

Define for h > 0, the modulus of continuity

$$\omega_n(h) := \sup\{|\log \Psi(c_n x) - \log \Psi(c_n y)| : |x-y| \le h, |x|, |y| \le c_m/c_n\}. \tag{10.20}$$

We split the estimation of $\omega_n(h)$ into two parts. Let $\ell = \ell(n)$, $n = 1, 2, 3, \ldots$ be a non–decreasing sequence of positive integers, with limit ∞, as $n \to \infty$. We have for

$$c_\ell \le c_n x \le c_n y \le c_m, \; y \le x + h,$$

$$|\log \Psi(c_n x) - \log \Psi(c_n y)| = |-\tfrac{1}{2} \log(\frac{T(c_n x)}{T(c_n y)}) + \log(\frac{Z(c_n x)}{Z(c_n y)})|$$

$$\le \tfrac{1}{2} |\int_{c_n x}^{c_n y} \frac{T'(t)}{T(t)} dt| + C_3 Q(c_n x)^{-1/2}$$

(by (10.19))

$$\le C_4\, T(c_m) \log(y/x) + C_4 Q(c_\ell)^{-1/2},$$

(by (3.42))

$$\le C_5\, n^{2/29}(h/x) + C_5\, \ell^{-27/58}$$

(by (3.25) and (3.35) and since $a_n \le c_n \le a_{2n}$, see (3.34))

$$\le C_6\, n^{2/29} c_n h + C_6\, \ell^{-27/58}, \tag{10.21}$$

for $n \ge n_1$, since $x^{-1} \le c_n/c_\ell \le c_n$, n large enough. Next, there exists $K > 0$ s.t. $Q'(x)/x$ is increasing in $[K, \infty)$ (by Lemma 3.1(b)). We may assume that $K \ge C_2$ where C_2 is as in (10.16) and (10.17). Then if

$$K \le c_n x \le c_n y \le c_\ell, \; y \le x + h,$$

we have

$$|\log \Psi(c_n x) - \log \Psi(c_n y)| = |\int_{c_n x}^{c_n y} \Psi'(t)/\Psi(t)\, dt|$$

$$\le C_7 \int_{c_n x}^{c_n y} \{Q'(t)/t\}^{1/2}\, T(t)^{1/2}\, dt$$

$$\le C_8\, c_n\, h\, \{T(c_\ell)\, Q'(c_\ell)/c_\ell\}^{1/2}$$

$$\leq C_9\, c_n\, h\, \ell^{16/29}, \tag{10.22}$$

by (3.25) and (3.30). Finally, if

$$0 \leq c_n x \leq c_n y \leq K,\ y \leq x + h,$$

then from (3.2),

$$0 \leq Q(c_n y) - Q(c_n x) \leq Q'(K) \int_{c_n x}^{c_n y} (\tfrac{t}{K})^{T(t)-1}\, dt$$

$$\leq C_{10}\, (c_n h)^{\min\{1,T(0+)\}}, \tag{10.23}$$

by the monotonicity of T, and some straightforward manipulations. Since $G_{Q/2}^{\pm 1}$ is continuously differentiable in [–K,K], it follows that $\Psi(c_n \cdot)$ admits a similar estimate to (10.23). Combining (10.21) to (10.23) yields

$$\omega_n(h) \leq C_{11}\{n^{2/29} c_n h + \ell^{-27/58} + c_n\, h\, \ell^{16/29} + (c_n h)^{\min\{1,T(0+)\}}\}, \tag{10.24}$$

$h > 0$, with C_{11} independent of h, n, and $\ell = \ell(n)$. The most difficult or dominant term on the right–hand side of (10.24) is the first. In view of Lemma 3.2(g), and in view of (10.11), it is not difficult to see that we can choose $\ell = \ell(n)$ to grow like a sufficiently small power of n, in such a way that for some $\delta_1 > 0$,

$$\omega_n(n^{-\rho}) \leq n^{-\delta_1},\ n \text{ large enough.} \tag{10.25}$$

By Jackson's Theorem, we can find polynomials $U_n(x)$ of degree at most n^ρ such that for some $0 < \delta < \delta_1$,

$$|\log \Psi(c_n x) - U_n(x)| \leq n^{-\delta},\ |x| \leq c_m/c_n. \tag{10.26}$$

Then

$$|U_n(x)| \leq C_{12} \log T(c_m) \leq C_{13} \log n,\ |x| \leq c_m/c_n. \tag{10.27}$$

Let $V_k(x)$ denote the (k+1)th partial sum of the Maclaurin series of exp(x). We recall (9.15). Choose $k := \langle 16C_{13} \log n\rangle + 1$, n large enough, and set

$$R_n(x) := V_k(U_n(x)),$$

a polynomial of degree at most $32\, C_{13}\, (\log n)\, n^\rho$. From (10.26) and (9.15), we obtain (10.12). Further, for $|x| \le c_m/c_n$,

$$R_n'(x) = V_k'(U_n(x))U_n'(x) = V_{k-1}(U_n(x))U_n'(x)$$
$$= \Psi(c_n x)\, O((\log n)^3\, n^{2\rho}),$$

by (10.27) and the Markov inequality. Then from (10.16), for $\frac{c_m}{c_n} \ge |x| \ge K/c_n$,

$$\frac{d}{dx}\{R_n(x)/\Psi(c_n x)\} = R_n'(x)/\Psi(c_n x) - \{R_n(x)/\Psi(c_n x)\}\,\{c_n \Psi'(c_n x)/\Psi(c_n x)\}$$
$$= O((\log n)^3\, n^{2\rho}) + O(c_n\, \{Q'(c_m)T(c_m)/c_m\}^{1/2})$$
$$= O((\log n)^3\, n^{2\rho}) + O(n^{16/29}) = O(n^{16/29}),$$

assuming as we can, that $\rho \le 1/4$, and using Lemma 3.3. For $\epsilon/c_n \le |x| \le K/c_n$, we see that

$$\{c_n \Psi'(c_n x)/\Psi(c_n x)\} = O(c_n) = o(n^{1/2}).$$

Hence (10.13). Of course, if Q' is continuous at 0, then we can allow $|x| \le c_m/c_n$.

Finally, the estimate (10.14) for the second derivative follows similarly, using (10.17). Although $R_n(x)$ has degree $\sim$ $(\log n)\, n^\rho$, rather than $O(n^\rho)$, we can always start with a slightly smaller ρ satisfying (10.11), and then can satisfy the degree restriction, namely degree$(R_n) \le n^\rho$. □

Proof of Theorem 10.1

Suppose first that $\kappa \in (-1,1)$. According to Theorem 9.1, we can find polynomials $P_n(u)$ of degree at most $2n - 2k_n - 2\langle n\Delta_n \rangle$ satisfying (9.3) to (9.9) with $H := G_{Q/2}$. Let us denote those satisfying (9.4) by P_n^- and (if $\kappa \in [0,1)$) those satisfying (9.5) by P_n^+. Here if $\kappa \in (-1,0)$, the argument below applies only to P_n^- as Theorem 9.1 does not yield a suitable P_n^+. To distinguish $\tau_n(u)$ of Theorem 9.1 from that of the present theorem, let

$$\hat{\tau}_n^{\pm}(u) := P_n^{\pm}(u)/G_{Q/2}^2(c_n u)\,(1 - u^2)^{-\kappa},\ u \in (-1,1).$$

Thus (9.6) to (9.9) hold for $\hat{\tau}_n^{\pm}$, while $\hat{\tau}_n^-$ satisfies (9.4) and $\hat{\tau}_n^+$ satisfies (9.5). Let $R_n(x)$ be the polynomials of Lemma 10.2 with m = n, of degree at most $n^{2/28}$, say, n large enough. Define for some fixed $\delta_1 > 0$,

$$S_n^{\pm}(x) := P_n^{\pm}(x)\, R_n^2(x)\, (1 + n^{-\delta_1})^{\pm 1},$$

and

$$\tau_n^{\pm}(x) := S_n^{\pm}(x) W^2(c_n x)\, (1 - x^2)^{-\kappa}.$$

Note that $S_n^{\pm}(x)$ is a polynomial of degree at most $2n - 2k_n - 2<n\Delta_n> + 2n^{2/28} \leq 2n - 2k_n$, n large enough. From (9.7), (10.1), and (10.12), for n large enough and $|u| \leq 1 - n^{-\delta}$,

$$\begin{aligned}\tau_n^{\pm}(u) &= \{P_n^{\pm}(u)/G^2_{Q/2}(c_n u)\,(1 - u^2)^{-\kappa}\}\,\{R_n(u)/\Psi(c_n u)\}^2\,(1 + n^{-\delta_1})^{\pm 1}\\ &= \hat{\tau}_n^{\pm}(u)\,\{R_n(u)/\Psi(c_n u)\}^2\,(1 + n^{-\delta_1})^{\pm 1} = 1 + O(n^{-\delta}),\end{aligned}$$

provided δ is small enough. Hence (10.4). Next, (9.4), and (10.12) show that for $|u| \leq 1$,

$$\begin{aligned}\tau_n^-(u) &= \hat{\tau}_n^-(u)\,\{1 + O(n^{-\delta})\}\,(1 + n^{-\delta_1})^{-1}\\ &\leq 1 + O(n^{-\delta}) - n^{-\delta_1/2} \leq 1,\end{aligned}$$

provided δ_1 is small enough. Hence (10.3). Similarly we may derive (10.2). The bound (10.6) follows directly from (9.9) and (10.12). Then (10.5) follows from (10.4) and (10.6). The derivative estimates follow easily from (10.13), (10.14) and (9.7).

Finally, we turn to the case $\kappa = 1$. Let us start with the polynomials $\hat{S}_n^{\pm}(t)$ of this theorem for $\kappa = 0$, so that

$$W^2(c_n t)\, \hat{S}_n^+(t) \geq 1 \geq W^2(c_n t)\, \hat{S}_n^-(t),\ t \in [-1,1],$$

amongst other things. Set

$$S_n^{\pm}(t) := (\hat{S}_n^{\pm}(t) + n^{-2\delta})(1 - t^2)(1 + n^{-\delta})^{\pm 1}.$$

This is positive in [−1,1], except for simple zeros at ± 1. Further,

$$S_n^-(t)W^2(c_nt)$$

$$= (1-t^2)\{\hat{S}_n^-(t)W^2(c_nt)(1+n^{-\delta})^{-1} + n^{-2\delta}\, W^2(c_nt)(1+n^{-\delta})^{-1}\}$$

$$\le (1-t^2)\ \{(1+n^{-\delta})^{-1} + n^{-2\delta}\, W^2(0)\ (1+n^{-\delta})^{-1}\}$$

$$\le (1-t^2)\ \{1 - \tfrac{1}{2}\,n^{-\delta} + n^{-2\delta}\, W^2(0)\} \le 1-t^2,$$

$t \in [-1,1]$, for n large enough. It is easier to show that

$$S_n^+(t)W^2(c_nt) \ge 1 - t^2,\ t \in [-1,1].$$

The inequalities (10.4) to (10.8) for $S_n^{\pm}$ follow from the corresponding inequalities for $\hat{S}_n^{\pm}$, possibly with a smaller δ. □

11 Explicit Extremal Polynomials, Christoffel Functions and Szegö Functions

In this section, we gather together some well known material on extremal and orthogonal polynomials, on Christoffel functions, and on Szegö functions. First, some explicit extremal polynomials:

Lemma 11.1

Let q be a positive integer, let S(x) be a polynomial of degree at most 2q, positive in [−1,1], except possibly for simple zeros at ± 1, and let

$$V(x) := \begin{cases} [(1-x^2)/S(x)]^{1/2}, & x \in (-1,1), \\ 0 & , x \in \mathbb{R}\backslash(-1,1). \end{cases} \tag{11.1}$$

Further, for $0 < p \leq \infty$, let

$$V_p(x) := \begin{cases} (1-x^2)^{-1/2p}\, V(x), & x \in [-1,1], \\ 0 & , x \in \mathbb{R}\backslash(-1,1), \end{cases} \tag{11.2}$$

and recall σ_p from (1.23).

(a) For $1 \leq p \leq \infty$, and $n \geq q$,

$$E_{np}(V_p) := \inf_{P \in \mathcal{P}_{n-1}} \| \{x^n - P(x)\}\, V_p(x) \|_{L_p[-1,1]}$$
$$= \sigma_p\, 2^{-n+1}\, G[V] = \sigma_p\, 2^{-n+1-1/p}\, G[V_p], \tag{11.3}$$

and for $0 < p < 1$,

$$E_{np}(V_p) \leq \sigma_p\, 2^{-n+1-1/p}\, G[V_p]. \tag{11.4}$$

(b) Let $1 \leq p \leq \infty$ and let $T_{np}(V_p, x)$ denote the monic polynomial of degree n satisfying

$$\|T_{np}(V_p,x)V_p(x)\|_{L_p[-1,1]} = E_{np}(V_p). \tag{11.5}$$

Then for $1 \le p \le \infty$ and $n \ge q$, $x = \cos \phi \in [-1,1]$, $z := e^{i\phi}$,

$$T_{np}(V_p,x) = 2^{-n} G[V] \{z^{-n} D^{-2}(V(\cos t);z) + z^n D^{-2}(V(\cos t);z^{-1})\}, \tag{11.6}$$

and

$$|T_{np}(V_p,x)V(x)| \le 2^{-n+1-1/p} G[V_p], \; x \in [-1,1]. \tag{11.7}$$

Further, for $u \in \mathbb{C}\backslash[-1,1]$, and recalling the definition (1.22) of $\varphi(\cdot)$,

$$|T_{np}(V_p,u)/\{2^{-n-1/p}G[V_p] \varphi(u)^n D^{-2}(V(\cos t);\varphi(u)^{-1})\} - 1|$$
$$\le |\varphi(u)|^{2q-2n-2}. \tag{11.8}$$

<u>Proof</u>

This is Theorem 13.1 in [14,p.111–112]. The result itself is due mainly to Bernstein and Szegö. □

Next, a comparison lemma in the plane for polynomials:

<u>Lemma 11.2</u>

Let $W:\mathbb{R} \to [0, \infty)$ be a continuous non–negative function such that for $n = 0, 1, 2, \ldots$, $x^n W(x) \in L_\infty(\mathbb{R})$. Assume further that there exists an increasing seqence $\{c_n\}_{n=1}^{\infty}$ of positive numbers such that for $n = 1, 2, 3, \ldots$, and $2 \le p \le \infty$, and $P \in \mathcal{P}_n$,

$$\|PW\|_{L_p(\mathbb{R})} \le 2 \, \|PW\|_{L_p(-c_n,c_n)}. \tag{11.9}$$

Let $n \ge 1$, and let

$$P(x) := Ax^n + \ldots \in \mathcal{P}_n, \; A \ne 0. \tag{11.10}$$

Fix $2 \le p \le \infty$, and let

$$r_n := A \, E_{np}(W). \tag{11.11}$$

Then for $z \in \mathbb{C}\backslash \{x_{jn}^{(p)}\}_{j=1}^{n}$,

$$|P(z)/p_{np}(W,z) - r_n| \le 2\,\{\|PW\|^2_{L_p(\mathbb{R})} - r_n^2\}^{1/2}/d_n(z), \tag{11.12}$$

$$|P(z)/(AT_{np}(w,z)) - 1| \le 2\{\|PW\|^2_{L_p(\mathbb{R})} r_n^{-2} - 1\}^{1/2}/d_n(z), \tag{11.13}$$

where

$$d_n(z) := \min_{1\le j\le n} |z/c_n - x_{jn}^{(p)}/c_n|. \tag{11.14}$$

__Proof__

See Theorem 15.1 in [14,p.128] and Lemma 15.2 in [14,p.132]. □

Next, we recall some standard facts about the Szegö function. Let $g:[-1,1] \to [0,\infty)$ satisfy the Szegö condition

$$\int_{-1}^{1} |\log g(x)|\ (1-x^2)^{-1/2}\,dx < \infty. \tag{11.15}$$

Then if $g(\cos t)$, as a function of t, satisfies a Lipschitz condition of some positive order near $t = \phi$, and is positive there, we have with the notation (1.20) and (1.21), for $D(g(\cos t);z)$ and for $\Gamma(\cdot\,;\cdot\,)$,

$$D(g(\cos t);e^{i\phi}) = g(\cos\phi)^{1/2}\exp(i\Gamma(g;\phi)), \tag{11.16}$$

see [30,p.279]. Further, if $g'(\cos\phi)$ and $\Gamma'(g;\phi)$ exist,

$$ie^{i\phi}D'(g(\cos t);e^{i\phi})/D(g(\cos t);e^{i\phi})$$
$$= -\frac{1}{2}\frac{g'(\cos\phi)}{g(\cos\phi)}\sin\phi + i\Gamma'(g;\phi), \tag{11.17}$$

and in particular,

$$\mathrm{Re}\,[e^{i\phi}D'(g(\cos t);e^{i\phi})/D(g(\cos t);e^{i\phi})] = \Gamma'(g;\phi). \tag{11.18}$$

__Lemma 11.3__

Let $g:[-1,1] \to (0,\infty)$ satisfy (11.15). Let

$$h(x) := \log g(x),\ x \in (-1,1). \tag{11.19}$$

(a) Then for $x = \cos\phi \in (-1,1)$, $\phi \in (0,\pi)$, such that the integral exists,

$$\Gamma(g;\phi) = \frac{(1-x^2)^{1/2}}{2\pi} \int_{-1}^{1} \frac{h(s)-h(x)}{s-x} (1 - s^2)^{-1/2} \, ds. \tag{11.20}$$

Furthermore, if g is even, then

$$\Gamma(g;\pi-\phi) = -\Gamma(g;\phi), \tag{11.21}$$

and, if defined, then

$$\Gamma(g;\pi/2) = 0. \tag{11.22}$$

(b) For $x = \cos \phi \in (-1,1)$,

$$|\Gamma(g;\phi)| \le \frac{(1-x^2)^{1/2}}{2\pi} \int_{-1}^{1} \left|\frac{h(s)-h(x)}{s-x}\right| (1 - s^2)^{-1/2} \, ds, \tag{11.23}$$

and

$$|D(g(\cos t);e^{i\phi}) - 1|$$

$$\le |h(x)| + \frac{(1-x^2)^{1/2}}{\pi} \int_{-1}^{1} \left|\frac{h(s)-h(x)}{s-x}\right| (1 - s^2)^{-1/2} \, ds, \tag{11.24}$$

provided this last right–hand side is bounded above by 2.

(c) If h' is continuous in $(-1,1)$, and $h'(t)(1 - t^2)^{1/2} \in L_1[-1,1]$, then for $x = \cos \phi \in (-1,1)$,

$$\Gamma'(g;\phi) = -\frac{PV}{2\pi} \int_{-1}^{1} \frac{h'(t)(1-t^2)^{1/2}}{t - x} \, dt, \tag{11.25}$$

$$= -\frac{1}{2\pi} \int_{-1}^{1} \frac{h'(t)(1-t^2) - h'(x)(1-x^2)}{t - x} (1-t^2)^{-1/2} \, dt, \tag{11.26}$$

provided $\Gamma'(g;\phi)$ exists and these integrals are finite (PV denotes principal value). If, in addition, g is even, then

$$\Gamma'(g;\phi) = -\frac{1}{\pi} \int_{0}^{1} \frac{th'(t)(1-t^2)-xh'(x)(1-x^2)}{t^2-x^2} (1-t^2)^{-1/2} dt. \tag{11.27}$$

(d) Suppose that $x = \cos \phi \in (-1,1)$ is fixed, that $0 < \epsilon < (1-|x|)/2$, and that h' is continuous in $(x-\epsilon,x+\epsilon)$. Let

$$\omega(\eta) := \sup\{|h'(t)-h'(x)| : t \in (x-\eta,x+\eta)\}, \ \eta \in [0,\epsilon]. \tag{11.28}$$

Then

$$|\Gamma'(g;\phi)| \leq \frac{1}{\epsilon\pi}\int_{-1}^{1} \left|\frac{h(s)-h(x)}{s-x}\right| (1-s^2)^{-1/2}\, ds + 2|h'(x)|\ \log(\frac{2e}{\epsilon})$$

$$+ (1-|x|)^{1/2}\int_{0}^{\epsilon} \omega(\eta)/\eta\ d\eta. \tag{11.29}$$

Proof

(a) For (11.20), see [30,p.299]. If also g is even, we deduce that for $x = \cos\phi$,

$$\Gamma(g;\phi) = \frac{x}{\pi}(1-x^2)^{1/2}\int_{0}^{1} \frac{h(s)-h(x)}{s^2-x^2}\, ds. \tag{11.30}$$

Then, since $\cos(\pi-\phi) = -\cos(\phi) = -x$, we obtain (11.21), and also (11.22).

(b) Firstly, (11.23) is immediate. Next,

$$D(g(\cos t);e^{i\phi}) - 1$$
$$= g(\cos\phi)^{1/2}\exp(i\Gamma(g;\phi)) - 1$$
$$= \{e^{h(x)/2} - 1\}\exp(i\Gamma(g;\phi)) + \{\exp(i\Gamma(g;\phi)) - 1\}.$$

Using the inequality

$$|e^u - 1| \leq 2|u|,\ |u| \leq 1, \tag{11.31}$$

and using (11.23), we obtain (11.24), provided the right–hand side there is bounded above by 2.

(c) Differentiating the right–hand side of (11.20) with respect to x, and using $\frac{dx}{d\phi} = -\sin\phi$, yields

$$\Gamma'(g;\phi) = \frac{x}{2\pi}\int_{-1}^{1} \frac{h(s)-h(x)}{s-x}(1-s^2)^{-1/2}\, ds$$

$$-\frac{(1-x^2)}{2\pi}\int_{-1}^{1} \frac{h(s)-h(x)-h'(x)(s-x)}{(s-x)^2}(1-s^2)^{-1/2}\, ds. \tag{11.32}$$

Noting the identities

$$PV\int_{-1}^{1} \frac{1}{s-a}(1-s^2)^{-1/2}\, ds = 0,\ a \in (-1,1), \tag{11.33}$$

and

$$\frac{1-x^2}{s-x} = \frac{1-s^2}{s-x} + s + x,$$

we obtain

$$\Gamma'(g;\phi) = \frac{x}{2\pi}\int_{-1}^{1} \frac{h(s)-h(x)}{s-x}(1-s^2)^{-1/2}\,ds$$

$$-\frac{(1-x^2)}{2\pi}\,PV\int_{-1}^{1} \frac{h(s)-h(x)}{(s-x)^2}(1-s^2)^{-1/2}\,ds$$

$$= \frac{PV}{2\pi}\int_{-1}^{1} \frac{h(s)-h(x)}{s-x}\left(\frac{x}{(1-s^2)^{1/2}} - \frac{(1-s^2)^{1/2}}{s-x} - \frac{s+x}{(1-s^2)^{1/2}}\right) ds$$

$$= \frac{PV}{2\pi}\int_{-1}^{1} [h(s)-h(x)]\left(\frac{-s}{(1-s^2)^{1/2}}\,\frac{1}{s-x} - \frac{(1-s^2)^{1/2}}{(s-x)^2}\right) ds$$

$$= \frac{PV}{2\pi}\int_{-1}^{1} [h(s)-h(x)]\,\frac{d}{ds}\left(\frac{(1-s^2)^{1/2}}{s-x}\right) ds$$

$$= \frac{PV}{2\pi}\left\{[h(s)-h(x)]\left(\frac{(1-s^2)^{1/2}}{s-x}\right)\Bigg|_{s=-1}^{s=1} - \int_{-1}^{1} h'(s)\frac{(1-s^2)^{1/2}}{s-x}\,ds\right\}$$

$$= -\frac{PV}{2\pi}\int_{-1}^{1} h'(s)\frac{(1-s^2)^{1/2}}{s-x}\,ds.$$

The integration by parts is easily justified, using the continuity of h' in (–1,1), the identity (11.33), and the convergence of the integral in (11.26). Thus, we have (11.25), and then (11.26) follows. If g is even, (11.27) also follows, using the fact that h' is odd.

(d) We use (11.32). If $|x-s| < \epsilon$, then for some ξ between x and s,

$$|h(s)-h(x)-h'(x)(s-x)| = |h'(\xi) - h'(x)|\,|s-x|$$

$$\le \omega(|s-x|)\,|s-x|.$$

Hence

$$\frac{1-x^2}{2\pi}\int_{|x-s|\le\epsilon} \frac{|h(s)-h(x)-h'(x)(s-x)|}{|s-x|^2}(1-s^2)^{-1/2}\,ds$$

$$\le \frac{(1-|x|)}{\pi(1-[|x|+\epsilon])^{1/2}}\int_{|x-s|\le\epsilon} \frac{\omega(|s-x|)}{|s-x|}\,ds$$

$$\leq \frac{\sqrt{2}\ (1-|x|)^{1/2}}{\pi}\ 2\int_0^{\epsilon} \omega(\eta)/\eta\ d\eta,$$

since

$$1 - [|x| + \epsilon] \geq (1 - |x|)/2.$$

Also,

$$\frac{1-x^2}{2\pi}\int_{|x-s|\geq\epsilon} \frac{|h(s)-h(x)-h'(x)(s-x)|}{|s-x|^2}\,(1-s^2)^{-1/2}\,ds$$

$$\leq \frac{1}{2\pi}\Big[\frac{1}{\epsilon}\int_{-1}^{1} \Big|\frac{h(s)-h(x)}{s-x}\Big|\ (1-s^2)^{-1/2}\,ds$$

$$+ (1-x^2)|h'(x)|\int_{|x-s|\geq\epsilon} |s-x|^{-1}\,(1-s^2)^{-1/2}\,ds\Big].$$

To estimate the second integral, we let

$$\mathscr{S} := \{s \in [-1,1]: |s-x| \geq \epsilon \text{ and } 1-|s| \leq (1-|x|)/2\},$$

and

$$\mathscr{T} := \{s \in [-1,1]: |s-x| \geq \epsilon \text{ and } 1-|s| > (1-|x|)/2\}.$$

If first, $s \in \mathscr{S}$, then

$$|s-x| \geq |s| - |x|$$

$$= (1-|x|) - (1-|s|) \geq (1-|x|)/2.$$

Hence,

$$(1-x^2)\int_{\mathscr{S}} |s-x|^{-1}\,(1-s^2)^{-1/2}\,ds$$

$$\leq \frac{2(1-|x|)}{(1-|x|)/2}\int_{\mathscr{S}} (1-s^2)^{-1/2}\,ds \leq 4\pi.$$

On the other hand,

$$(1-x^2)\int_{\mathscr{T}} |s-x|^{-1}\,(1-s^2)^{-1/2}\,ds$$

$$\leq \frac{2(1-|x|)}{((1-|x|)/2)^{1/2}}\int_{\mathscr{T}} |s-x|^{-1}\,ds \leq 4\sqrt{2}\,\log(2/\epsilon).$$

Then (11.29) follows from (11.32). □

Next, explicit orthogonal polynomials and Christoffel functions for special weights.

Lemma 11.4

Let Q(x) be a polynomial of degree at most ν, positive in [−1,1], except possibly for simple zeros at ± 1. Let

$$w(x) := (1 - x^2)^{1/2}/Q(x), \; x \in (-1,1), \tag{11.34}$$

and let

$$\mu'(\theta) := w(\cos\,\theta)|\sin\,\theta|, \; \theta \in [-\pi,\pi]. \tag{11.35}$$

Then for $n > \nu/2 - 1$, $x = \cos\,\phi \in (-1,1)$, $z := e^{i\phi}$,

$$p_n(w,x) = (2\pi)^{-1/2}\,\{z^n/D(\mu';z^{-1}) + z^{-n}/D(\mu';z)\}. \tag{11.36}$$

Further, for $n > \nu/2$,

$$\pi\lambda_n^{-1}(w,x)\,(1-x^2)^{1/2}\,w(x) = n - \tfrac{1}{2} +$$

$$\{2(1-x^2)^{1/2}\}^{-1}\mathrm{Im}[z^{2n-1}D(\mu';z)/\overline{D(\mu';z)}] + \mathrm{Re}[zD'(\mu';z)/D(\mu';z)], \tag{11.37}$$

$$= n - \tfrac{1}{2} + \{2\sin\,\phi\}^{-1}\,\sin((2n-1)\phi + 2\Gamma(g;\phi)) + \Gamma'(g;\phi), \tag{11.38}$$

where

$$g(x) := w(x)(1-x^2)^{1/2} = \mu'(\mathrm{arc\,cos}\,x), \; x \in [-1,1]. \tag{11.39}$$

Proof

For the case where $Q(x) = (1 - x^2)S(x)$, $S(x) > 0$ in [−1,1], (11.37) appears in [4,p.251], but with some slight errors. The latter are repeated in a different form in [4,p.269,Problem 13]. We sketch the details. Suppose first that Q(x) is positive in [−1,1], and let

$$\hat{w}(x) := w(x)/(1 - x^2) = 1/\{Q(x)\,(1 - x^2)^{1/2}\}, \; x \in [-1,1],$$

and

$$\hat{\mu}'(\theta) := \hat{w}(\cos\theta)|\sin\theta| = 1/Q(\cos\theta),\ \theta \in [-\pi,\pi].$$

Then, recalling the notation (1.24) to (1.25),

$$\phi_n(d\hat{\mu},z) = z^n/\overline{D(d\hat{\mu};1/\overline{z})},\ n \geq \nu,\ |z| \geq 1, \tag{11.40}$$

see [30,p.289,eqn.(11.2.2) and p.300,eqn.(12.3.1)] or [4,p.224,eqn.(4.16)]. Further,

$$\kappa_n(d\hat{\mu}) = 1/D(d\hat{\mu};0),\ n \geq \nu, \tag{11.41}$$

and since $\phi_n(d\hat{\mu},z) = z^{n-\nu}\ \phi_\nu(d\hat{\mu},z)$, $n > \nu$, we obtain

$$\phi_n(d\hat{\mu},0) = 0,\ n > \nu. \tag{11.42}$$

Then using eqn. (11.5.2) in [30,p.294] or eqn. (1.11) in [4,p.192], we obtain (11.36), after some straightforward manipulations, and after using (10.2.13) in [30,p.277]. Also, then

$$\gamma_n(w) = 2^n(2\pi)^{-1/2}/D(d\hat{\mu};0),\ 2n+2 > \nu. \tag{11.43}$$

Since

$$\lambda_n^{-1}(w,x) = \frac{\gamma_{n-1}(w)}{\gamma_n(w)}\{p_n'(w,x)\ p_{n-1}(w,x) - p_n(w,x)\ p_{n-1}'(w,x)\}$$

(cf. [4,p.252], but the sign there is wrong), we can use (11.36) and (11.43) to calculate $\lambda_n(w,x)$ by a straightforward, but lengthy, calculation. We spare the reader the details, but will furnish the pages of calculation on request. The result of the calculation is (11.37), and (11.38) then follows from (11.16) and (11.18). Finally, in the case where Q has simple zeros at 1 and/or –1, one can approximate the given Q, in a suitable sense, by a sequence of polynomials of the same degree, positive in [–1,1]. □

<u>Lemma 11.5</u>

Let $W := e^{-Q}$, where Q is even and continuous in $\mathbb{R}$, Q'' exists and is continuous in $(0,\infty)$, while Q'(x) is positive there, and xQ'(x) is strictly increasing in $(0,\infty)$, with

(4.1) and (4.2) holding for some $\eta \in (0,1)$.

(a) Let $n \geq 1$, let $R > 0$, and let $\mu_{n,R}$ be defined by (4.3) to (4.5). Let

$$F_R(x) := W(Rx)\,(1-x^2)^{1/4}, \; x \in (-1,1). \tag{11.44}$$

Then for $x = \cos\phi \in (-1,1)\backslash\{0\}$,

$$n + \Gamma'(F_R^2;\phi) = n\pi \sin\phi\, \mu_{n,R}(\cos\phi) + 1/2, \tag{11.45}$$

and for $x = \cos\phi \in (-1,1)$,

$$n\phi + \Gamma(F_R^2;\phi) = n\pi \int_x^1 \mu_{n,R}(t)\,dt + \phi/2 - \pi/4. \tag{11.46}$$

If Q' satisfies a Lipschitz condition of some positive order at 0, then (11.45) remains valid for $x = 0$.

(b) Let $R_j > 0$ and $x_j = \cos\phi_j \in (-1,1)$, $j = 1, 2$, satisfy

$$R_1x_1 = R_2x_2, \tag{11.47}$$

and assume that

$$\beta := R_1/R_2 \in [1/2,1). \tag{11.48}$$

Then

$$\begin{aligned} &|\Gamma'(F_{R_1}^2;\phi_1) - \Gamma'(F_{R_2}^2;\phi_2)| \\ &\leq \frac{4}{\pi}(1-\beta)\int_0^\beta \left|\frac{R_2sQ'(R_2s)-R_2x_2Q'(R_2x_2)}{s^2-x_2^2}\right| (1-s^2)^{-1/2}\,ds \\ &+ \frac{2}{\pi}\int_\beta^1 \left|\frac{R_2sQ'(R_2s)-R_2x_2Q'(R_2x_2)}{s^2-x_2^2}\right| (1-s^2)^{1/2}\,ds. \end{aligned} \tag{11.49}$$

<u>Proof</u>

(a) Let $h(x) := \log F_R^2(x)$, $x \in (-1,1)$. We see that

$$h(x) = -2Q(Rx) + \frac{1}{2}\log(1-x^2),$$

and hence that

$$(1-x^2)xh'(x) = -2RxQ'(Rx)\,(1-x^2) - x^2.$$

Then from (11.27), for $x = \cos\phi \in (-1,1)\backslash\{0\}$,

$$\Gamma'(F_R^2;\phi)$$

$$= \frac{2}{\pi} \int_0^1 \frac{RtQ'(Rt)(1-t^2) - RxQ'(Rx)(1-x^2)}{t^2 - x^2} (1-t^2)^{-1/2} \, dt$$

$$+ \frac{1}{\pi} \int_0^1 \frac{t^2 - x^2}{t^2 - x^2} (1-t^2)^{-1/2} \, dt$$

$$= \frac{2}{\pi} (1 - x^2) \int_0^1 \frac{RtQ'(Rt) - RxQ'(Rx)}{t^2 - x^2} (1-t^2)^{-1/2} \, dt$$

$$- \frac{2}{\pi} \int_0^1 RtQ'(Rt)(1-t^2)^{-1/2} \, dt + \frac{1}{2} \tag{11.50}$$

$$= n\pi (1 - x^2)^{1/2} \nu_{n,R}(x) + n(B_{n,R} - 1) + \frac{1}{2},$$

(by (4.4) and (4.5))

$$= n\pi (1 - x^2)^{1/2} \mu_{n,R}(x) - n + \frac{1}{2},$$

by (4.3). Hence (11.45). Since $\Gamma(F_R^2;\pi/2) = 0$, (see (11.22)), we obtain

$$n\phi + \Gamma(F_R^2;\phi) - n\pi/2$$

$$= \int_{\pi/2}^{\phi} (n + \Gamma'(F_R^2;\theta)) \, d\theta$$

$$= - n\pi \int_0^x \mu_{n,R}(t) \, dt + \phi/2 - \pi/4,$$

by (11.45). Since

$$\int_0^1 \mu_{n,R}(t) \, dt = \frac{1}{2},$$

we obtain (11.46).

(b) Using the identity

$$\frac{1-x^2}{t^2-x^2} (1-t^2)^{-1/2} = \frac{(1-t^2)^{1/2}}{t^2-x^2} + (1-t^2)^{-1/2},$$

in (11.50) leads to

$$\Gamma'(F^2_{R_1};\phi_1) = \frac{2}{\pi}\int_0^1 \frac{R_1 tQ'(R_1 t) - R_1 x_1 Q'(R_1 x_1)}{t^2 - x_1^2}(1-t^2)^{1/2}\,dt$$

$$- R_1 x_1 Q'(R_1 x_1) + \frac{1}{2}.$$

Making the substitutions

$$R_1 t = R_2 s;\ R_1 x_1 = R_2 x_2$$

in this integral yields,

$$\Gamma'(F^2_{R_1};\phi_1)$$

$$= \frac{2}{\pi}\int_0^\beta \frac{R_2 sQ'(R_2 s) - R_2 x_2 Q'(R_2 x_2)}{\beta^{-2}(s^2 - x_2^2)}(1-(\beta^{-1}s)^2)^{1/2}\,\beta^{-1}\,ds$$

$$- R_2 x_2 Q'(R_2 x_2) + \frac{1}{2}.$$

Then

$$\Gamma'(F^2_{R_1};\phi_1) - \Gamma'(F^2_{R_2};\phi_2)$$

$$= \frac{2}{\pi}\int_0^\beta \frac{R_2 sQ'(R_2 s) - R_2 x_2 Q'(R_2 x_2)}{s^2 - x_2^2}\{\beta(1-(\beta^{-1}s)^2)^{1/2} - (1-s^2)^{1/2}\}\,ds$$

$$- \frac{2}{\pi}\int_\beta^1 \frac{R_2 sQ'(R_2 s) - R_2 x_2 Q'(R_2 x_2)}{s^2 - x_2^2}(1 - s^2)^{1/2}\,ds.$$

Finally,

$$|\beta(1-(\beta^{-1}s)^2)^{1/2} - (1-s^2)^{1/2}|$$

$$= |(\beta^2-s^2)^{1/2} - (1-s^2)^{1/2}|\,\frac{(\beta^2-s^2)^{1/2} + (1-s^2)^{1/2}}{(\beta^2-s^2)^{1/2} + (1-s^2)^{1/2}}$$

$$= \frac{1 - \beta^2}{(\beta^2-s^2)^{1/2} + (1-s^2)^{1/2}} \le \frac{2(1-\beta)}{(1-s^2)^{1/2}}. \qquad \square$$

Next, a comparison lemma for orthogonal polynomials, contained in [30,pp.306–8], though not explicitly stated.

Lemma 11.6

Let f, f_1, f_2 be non–negative measurable functions on $[-\pi,\pi]$ such that

$$f_1(\theta) \le f(\theta) \le f_2(\theta),\ \theta \in [-\pi,\pi], \tag{11.51}$$

and assume that

$$f_2(\theta) = 1/g_2(\theta), \tag{11.52}$$

$$f_1(\theta) = (\sin\,\theta)^{\chi}/g_1(\theta), \tag{11.53}$$

where g_1 and g_2 are trigonometric polynomials of degree < m, positive in $[-\pi,\pi]$, and χ is a non–negative even integer. Let $a := e^{i\phi}$, some $\phi \in (0,\pi)$, and assume that

$$f_2(\phi) = f_1(\phi). \tag{11.54}$$

Then

$$\begin{aligned}|\,\phi_m(f,a)\ /\{a^m/\overline{D(f_2;a)}\} - 1| &\le |\,D(f_2;0)/D(f_1;0) - 1| \\ &+ |\,\{D(f_2;0)/D(f_1;0)\}^2 - 1|^{1/2}\ \{\frac{1}{4\pi}\int_{-\pi}^{\pi}\frac{\log(f_2/f_1)(t)}{(\sin(\frac{t-\phi}{2}))^2}dt\}^{1/2}.\end{aligned} \tag{11.55}$$

Proof

We extract a few of the details from Szegö's argument. Defining

$$s_m(f,a,z) = \sum_{j=0}^{m} \overline{\phi_j(f,a)}\ \phi_j(f,z),$$

Szegö [30,p.306,ineq.(12.6.7)] shows that

$$|\,\kappa_m(f)\ \phi_m(f,a) - \kappa_m(f_2)\ \phi_m(f_2,a)|^2$$

$$\le \{\kappa_m(f_1)^2 - \kappa_m(f_2)^2\}\ \{s_m(f_1,a,a) - s_m(f_2,a,a)\}, \tag{11.56}$$

this being a consequence of (11.51) only. Next, also as a consequence of (11.51), and of (11.41),

$$1/D(f_2,0) = \kappa_m(f_2) \le \kappa_m(f) \le \kappa_m(f_1) \le 1/D(f_1;0). \tag{11.57}$$

Further, as g_2 is of degree < m, (see (11.40)),

$$|\,\phi_m(f_2,a)| = 1/|\,D(f_2;a)| = f_2(\phi)^{-1/2}. \tag{11.58}$$

Then (11.56) to (11.58) yield

$$|\phi_m(f,a)/\phi_m(f_2,a) - 1| \le |D(f_2;0)/D(f_1;0) - 1|$$
$$+ |\{D(f_2;0)/D(f_1;0)\}^2 - 1|^{1/2} f_2(\phi)^{1/2} \{s_m(f_1,a,a) - s_m(f_2,a,a)\}^{1/2}. \tag{11.59}$$

Next [30,p.307,eqn.(12.6.11')],

$$s_m(f_2,a,a) = f_2(\phi)^{-1}\{m + 1 + 2 \text{ Re } [aD'(f_2;a)/D(f_2;a)]\},$$

while [30,p.308,eqn.(12.6.18)]

$$s_m(f_1,a,a) \le f_1(\phi)^{-1}\{m + 1 + 2 \text{ Re } [aD'(f_1;a)/D(f_1;a)]\}.$$

In view of (11.54), we then obtain

$$f_2(\phi)\ \{s_m(f_1,a,a) - s_m(f_2,a,a)\}$$
$$\le 2 \text{ Re } [aD'(f_1;a)/D(f_1;a) - aD'(f_2;a)/D(f_2;a)]. \tag{11.60}$$

Finally, it is an easy consequence of (11.51) and (11.54) that

$$f_1'(\phi) = f_2'(\phi).$$

Then (11.59), (11.60), (11.40) and the identity (10.4.23) in [30,p.284] yield the lemma.

□

We shall need two more comparison lemmas, the first typical of those used in the "relativized" Szegö theory [17,28].

<u>Lemma 11.7</u>

Let f, $g \in L_1[-\pi,\pi]$ be non–negative functions such that for some $0 < \eta < 1$,

$$(1 + \eta)^{-1} \le f(\theta)/g(\theta) \le 1 + \eta,\ \theta \in [-\pi,\pi]. \tag{11.61}$$

Further, assume that f is even. Then given $m \ge 2$, $\phi \in [0,\pi]$, $z := e^{i\phi}$ and $0 < \epsilon < 1$,

$$|\phi_m(g,z)/\phi_m(f,z) - 1| \le 4\eta/\epsilon + 2 \max_{|\sin(\frac{\theta-\phi}{2})|\le\epsilon} |\phi_m(f,e^{i\theta})f(\theta)^{1/2}|$$

$$\times \Big\{\frac{1}{2\pi}\int_{|\sin(\frac{\theta-\phi}{2})|\le\epsilon} \Big|\frac{(f/g)(\theta)-(f/g)(\phi)}{\sin((\theta-\phi)/2)}\Big|^2 d\theta\Big\}^{1/2}. \tag{11.62}$$

<u>Proof</u>

Let $s_m(f,z,a)$ be as in the previous lemma. Then [30,p.293]

$$s_{m-1}(f,z,u) = \frac{\overline{\phi_m^*(f,z)}\,\phi_m^*(f,u) - \overline{\phi_m(f,z)}\,\phi_m(f,u)}{1 - \overline{z}u},$$

where $\phi_m^*(f,z) := z^m\,\overline{\phi_m(f,1/\overline{z})}$. Then [30,p.293], writing $u := e^{i\theta}$,

$$\overline{\phi_m(g,z)} - \overline{\phi_m(f,z)}\,\frac{\kappa_m(g)}{\kappa_m(f)}$$

$$= \frac{1}{2\pi}\int_{-\pi}^{\pi} s_{m-1}(f,z,u)\,[\overline{\phi_m(g,u)} - \overline{\phi_m(f,u)}\,\frac{\kappa_m(g)}{\kappa_m(f)}]\,f(\theta)\,d\theta$$

$$= \frac{1}{2\pi}\int_{-\pi}^{\pi} s_{m-1}(f,z,u)\,[\overline{\phi_m(g,u)}]\,\{f(\theta) - \frac{f(\phi)}{g(\phi)}g(\theta)\}\,d\theta$$

$$= \overline{\phi_m^*(f,z)}\,\frac{1}{2\pi}\int_{-\pi}^{\pi} \phi_m^*(f,u)\,\overline{\phi_m(g,u)}\,\Big\{\frac{f(\theta) - \frac{f(\phi)}{g(\phi)}g(\theta)}{\overline{z}(z-u)}\Big\}\,d\theta$$

$$- \overline{\phi_m(f,z)}\,\frac{1}{2\pi}\int_{-\pi}^{\pi} \phi_m(f,u)\,\overline{\phi_m(g,u)}\,\Big\{\frac{f(\theta) - \frac{f(\phi)}{g(\phi)}g(\theta)}{\overline{z}(z-u)}\Big\}\,d\theta.$$

Since f is even, $|\phi_m^*(f,z)| = |\phi_m(f,z)|$, and hence

$$|\phi_m(g,z)/\phi_m(f,z) - \kappa_m(g)/\kappa_m(f)|$$

$$\le \frac{1}{\pi}\int_{-\pi}^{\pi} |\phi_m(f,u)\phi_m(g,u)|\,\Big|\frac{f(\theta) - \frac{f(\phi)}{g(\phi)}g(\theta)}{2\sin((\theta-\phi)/2)}\Big|\,d\theta \tag{11.63}$$

$$\le (1+\eta)^{1/2}\,\frac{1}{2\pi}\int_{-\pi}^{\pi} |\phi_m(f,u)f(\theta)^{1/2}|\,|\phi_m(g,u)g(\theta)^{1/2}|\,\Big|\frac{(f/g)(\theta)-(f/g)(\phi)}{\sin((\theta-\phi)/2)}\Big|\,d\theta$$

$$\le \sqrt{2}\max_{|\sin(\frac{\theta-\phi}{2})|\le\epsilon} |\phi_m(f,u)f(\theta)^{1/2}|\,\Big[\frac{1}{2\pi}\int_{|\sin(\frac{\theta-\phi}{2})|\le\epsilon} \Big|\frac{(f/g)(\theta)-(f/g)(\phi)}{\sin((\theta-\phi)/2)}\Big|^2\,d\theta\Big]^{1/2}$$

$$+ \sqrt{2} \max_{|\sin(\frac{\theta-\phi}{2})| \geq \epsilon} \left| \frac{(f/g)(\theta) - (f/g)(\phi)}{\sin((\theta-\phi)/2)} \right|,$$

by the Cauchy–Schwarz inequality and orthonormality. Finally, from (11.61),

$$|\kappa_m(g)/\kappa_m(f) - 1| \leq \eta. \tag{11.64}$$

Applying (11.61) once more, yields (11.62), since $\epsilon < 1$. □

We need one more (apparently new) comparison lemma. The point of it is that $\phi_m(f,z)/\phi_m(f,z)$ may be close to 1, even when $(f/g)'$ is large, provided that f/g is close enough to 1. The factor β below can apparently be removed with extra effort, but we omit this, as we shall not need it.

<u>Lemma 11.8</u>

Let f, g $\in L_1[-\pi,\pi]$ be even, non–negative, 2π–periodic functions such that for some $0 < \eta < 1$, (11.61) holds. Let $[a,b] \subset [0,\pi]$, let $(f/g)'$ exist in [a,b], and let

$$|(f/g)'(\theta)| \leq \nu, \ \theta \in [a,b]. \tag{11.65}$$

Let $m \geq 1$ be fixed, and let $M \geq 1$ satisfy

$$\sup_{\theta \in [a,b]} |\phi_m(f,e^{i\theta}) \, f(\theta)^{1/2}| \leq M. \tag{11.66}$$

For $\phi \in (0,\pi)$, let

$$\beta := \beta(\phi) := \cos(\tfrac{\pi+\phi}{4}). \tag{11.67}$$

Assume also that $j \geq 2$ and

$$0 < \epsilon < \min\{1, (b-a)/(2j)\}, \tag{11.68}$$

while

$$\eta \leq \epsilon \min\{1,\nu,\beta\}. \tag{11.69}$$

Let

$$B := 6M^2 \frac{\eta}{\beta} \log (7\nu/\eta). \tag{11.70}$$

Then, if also $\phi \in [a + j\epsilon, b - j\epsilon]$, we have

$$|\phi_m(g,e^{i\phi})/\phi_m(f,e^{i\phi}) - 1|$$
$$\leq \frac{10\eta}{\beta\epsilon} + 16j\ B \max\{1,B\}^{j-2} + 8\ \frac{M}{\beta}\ B^{j-1}\ (\eta\nu)^{1/2}. \tag{11.71}$$

<u>Proof</u>

From (11.61),

$$|(f/g)(\theta) - 1| \leq \eta,\ \theta \in [-\pi,\pi]. \tag{11.72}$$

We begin by estimating for $\phi \in [a+\epsilon, b-\epsilon]$, and for $k = 1, 2$,

$$J^{(k)} := \int_{-\pi}^{\pi} \left|\frac{(f/g)(\theta)-(f/g)(\phi)}{\theta-\phi}\right|^k d\theta$$
$$\leq \int_{|\theta-\phi|\geq\eta/\nu} (2\eta)^k|\theta-\phi|^{-k}\ d\theta + \int_{|\theta-\phi|<\eta/\nu} \nu^k\ d\theta$$
$$\leq 2^{k+1}\ \eta^k \left\{\begin{array}{ll} \log(2\pi\nu/\eta), & k=1 \\ (\eta/\nu)^{-1}, & k=2 \end{array}\right\} + 2\ \nu^k\ (\eta/\nu)$$
$$\leq \left\{\begin{array}{ll} 6\eta\ \log\ (7\nu/\eta), & k=1. \\ 10\nu\eta, & k=2. \end{array}\right. \tag{11.73}$$

Here we also used the fact that $\eta/\nu \leq 1$ (see (11.69)). Note that the dependence of $J^{(1)}$ on the size of ν is rather weak, while that of $J^{(2)}$ on ν is much stronger – this is the crux of the lemma. Next, recall from (11.64) that

$$|\kappa_m(g)/\kappa_m(f) - 1| \leq \eta. \tag{11.74}$$

Now let us set for $k = 0, 1, 2, \ldots j$,

$$I(k) := [a+k\epsilon, b-k\epsilon],$$

and for $k = 1, 2, \ldots j$,

$$N_k := \sup_{\phi\in I(k)} |\phi_m(g,e^{i\phi})\ g(\phi)^{1/2}|. \tag{11.75}$$

Recall the inequality

$$\sin s \geq \frac{2}{\pi} s,\ s \in [0,\tfrac{\pi}{2}],$$

which yields for $\theta \in [-\pi,\pi]$,

$$|\sin(\tfrac{\theta-\phi}{2})| = 2\,|\sin(\tfrac{\theta-\phi}{4})|\;|\cos(\tfrac{\theta-\phi}{4})|$$

$$\geq \pi^{-1}\,|\theta-\phi|\;\cos(\tfrac{\pi+\phi}{4}) = \tfrac{\beta}{\pi}\,|\theta-\phi|. \tag{11.76}$$

From (11.63), (11.74) and (11.76), for $\phi \in I(k)$, some $k \geq 1$,

$$\Delta(\phi) := |\,\phi_m(g,e^{i\phi})/\phi_m(f,e^{i\phi}) - 1|$$

$$\leq \eta + \frac{1}{2\beta}\int_{-\pi}^{\pi} |\,\phi_m(f,u)\phi_m(g,u)|\;g(\theta)\;\left|\frac{(f/g)(\theta)-(f/g)(\phi)}{\theta-\phi}\right|\;d\theta$$

$$(u = e^{i\theta})$$

$$\leq \eta + \frac{M}{2\beta}(1+\eta)^{1/2}\int_{I(k-1)} |\,\phi_m(g,u)g(\theta)^{1/2}|\left|\frac{(f/g)(\theta)-(f/g)(\phi)}{\theta-\phi}\right|\;d\theta$$

$$+\frac{\eta}{\beta\epsilon}(1+\eta)^{1/2}\int_{[-\pi,\pi]\backslash I(k-1)} |\,\phi_m(f,u)f(\theta)^{1/2}|\,|\,\phi_m(g,u)g(\theta)^{1/2}|\;d\theta. \tag{11.77}$$

Applying the Cauchy–Schwarz inequality in both integrals in the right–hand side of (11.77), with $k = 1$, yields, for $\phi \in [a+\epsilon,b-\epsilon]$, and since $\eta \leq 1$,

$$\Delta(\phi) \leq \eta + \frac{M}{\beta}\{\pi\,J^{(2)}\}^{1/2} + \frac{\eta}{\beta\epsilon}\sqrt{2}\;2\pi. \tag{11.78}$$

Applying (11.75) to the first integral in (11.77), and the Cauchy–Schwarz inequality in the second, yields for $j \geq k \geq 2$, and $\phi \in I(k)$,

$$\Delta(\phi) \leq \eta + \frac{M}{\beta}N_{k-1}\,J^{(1)}/\sqrt{2} + \frac{\eta}{\beta\epsilon}\sqrt{2}\;2\pi. \tag{11.79}$$

But for $\phi \in I(k)$,

$$|\,\phi_m(g,e^{i\phi})g(\phi)^{1/2}|$$

$$= |\,\{\phi_m(g,e^{i\phi})/\phi_m(f,e^{i\phi}) - 1\} + 1|\;|\,\phi_m(f,e^{i\phi})|\,g(\phi)^{1/2}$$

$$\leq (\Delta(\phi) + 1)\,M\sqrt{2} \tag{11.80}$$

$$\leq M\sqrt{2}\;\{\eta + \frac{M}{\beta}N_{k-1}\frac{6}{\sqrt{2}}\,\eta\log(7\nu/\eta) + \frac{\eta}{\beta\epsilon}\sqrt{2}\;2\pi + 1\},$$

by (11.73) and (11.79). Hence, recalling (11.70) and from (11.69) that $\eta/(\beta\epsilon) \leq 1$, we obtain

$$N_k \leq A + B\, N_{k-1},\ k = 2, 3, \dots j, \tag{11.81}$$

where

$$A := 16\, M.$$

Applying (11.81) repeatedly now yields

$$\begin{aligned} N_{j-1} &\leq A + B\, N_{j-2} \\ &\leq A + AB + B^2\, N_{j-3} \\ &\leq \dots \\ &\leq A + AB + AB^2 + \dots + AB^{j-3} + B^{j-2}\, N_1. \end{aligned}$$

But by (11.80),

$$\begin{aligned} N_1 &\leq \sup_{\phi \in [a+\epsilon, b-\epsilon]} (\Delta(\phi) + 1)\, M\sqrt{2} \\ &\leq \{\eta + \tfrac{M}{\beta}\, [10\pi\, \nu\eta]^{1/2} + \tfrac{\eta}{\beta\epsilon}\sqrt{2}\, 2\pi + 1\}\, M\sqrt{2} \\ &\qquad \text{(by (11.73) and (11.78))} \\ &\leq A + \tfrac{8}{\beta}\, M^2\, (\nu\eta)^{1/2}. \end{aligned}$$

Hence

$$N_{j-1} \leq A + AB + AB^2 + \dots AB^{j-3} + AB^{j-2} + \tfrac{8}{\beta}\, M^2 (\nu\eta)^{1/2} B^{j-2}.$$

Finally, substituting into (11.79) with k = j, yields for $\phi \in [a+j\epsilon, b-j\epsilon]$,

$$\begin{aligned} \Delta(\phi) &\leq \eta + \tfrac{M}{\beta}\, [Aj \max\{1,B\}^{j-2} + \tfrac{8}{\beta}\, M^2 (\nu\eta)^{1/2} B^{j-2}]\, 6\eta \log(7\nu/\eta) \\ &\quad + \tfrac{\eta}{\beta\epsilon}\sqrt{2}\, 2\pi \\ &\leq \tfrac{10\eta}{\beta\epsilon} + 16j\, B \max\{1,B\}^{j-2} + \tfrac{8}{\beta}\, M\, (\nu\eta)^{1/2} B^{j-1}, \end{aligned}$$

by (11.70) and definition of A. Hence (11.71). □

12 Extremal Errors and Extremal Polynomials

In this section, we establish the asymptotic behaviour of the extremal errors $E_{np}(W)$, and the behaviour of the extremal polynomials in the plane for $2 \leq p \leq \infty$. In particular this includes the proof of most of Theorem 2.2.

<u>Theorem 12.1</u>

Let $W \in SE(3)$. Then given $1 \leq p \leq \infty$, there exists $\delta > 0$ such that

$$E_{np}(W) = 2\sigma_p \left(\frac{a_n}{2}\right)^{n+1/p} G[W(a_n x)] \{1 + O(n^{-\delta})\}, \; n \to \infty. \tag{12.1}$$

If $0 < p < 1$, there exists $\delta > 0$ such that

$$E_{np}(W) \leq 2\sigma_p \left(\frac{a_n}{2}\right)^{n+1/p} G[W(a_n x)] \{1 + O(n^{-\delta})\}, \; n \to \infty. \tag{12.2}$$

<u>Proof</u>

We first establish an asymptotic upper bound in the case $0 < p \leq \infty$. Let $\{c_n\}_{n=1}^{\infty}$ be as in (5.6) of Theorem 5.2 with K large enough and S = 2. Then by (5.8) of Theorem 5.2,

$$E_{np}(W) \leq (1 + n^{-2}) \inf_{P \in \mathcal{P}_{n-1}} \| \{t^n - P(t)\} W(t) \|_{L_p[-c_n, c_n]}$$

$$= (1 + n^{-2}) c_n^{n+1/p} \inf_{P \in \mathcal{P}_{n-1}} \| \{t^n - P(t)\} W(c_n t) \|_{L_p[-1,1]}.$$

At least in the case $1 \leq p \leq \infty$, Theorem 10.1 (with $\kappa := 1 - 1/p$) guarantees the existence of polynomials $S_n^-(t)$ of degree at most $2n-4$ (or even of degree at most $2n-4-n^{1/4}$), positive in $[-1,1]$, except possibly for simple zeros at ± 1, s.t

$$W^2(c_n t) \leq (1 - t^2)^{1-1/p} / S_n^-(t) =: V_{p,n}(t)^2, \; t \in [-1,1], \tag{12.3}$$

and for some $\delta > 0$,

$$\int_{-1}^{1} |\log\{W^2(c_n t)/V_{p,n}^2(t)\}| \; (1-t^2)^{-1/2}\, dt = O(n^{-\delta}). \tag{12.4}$$

In the case $0 < p < 1$, we start with the polynomials S_n^- above for $p = 1$, and multiply by suitable polynomials of degree $<n^{1/4}>$. In fact since $1 - 1/p < 0$, the partial sums of the Maclaurin series of $(1 - t^2)^{1-1/p}$ will do, or we can multiply by the polynomials of Lemma 9.3 and can use (9.19) to (9.21). We leave the straightforward details to the reader. Thus for $0 < p \le \infty$, we can construct polynomials $S_n^-(t)$ of degree at most $2n-4$, positive in $[-1,1]$, except possibly for simple zeros at ± 1, satisfying (12.3) and (12.4).

Then for n large enough, (12.3) yields

$$\begin{aligned} E_{np}(W) &\le (1 + n^{-2})\, c_n^{n+1/p}\, E_{np}(V_{p,n}) \\ &\le (1 + n^{-2})\, c_n^{n+1/p}\, 2\sigma_p\, 2^{-n-1/p}\, G[V_{p,n}] \quad \text{(by (11.3) and (11.4))} \\ &= (1 + n^{-2})\, 2\sigma_p\, (c_n/2)^{n+1/p}\, G[W(c_n t)]\, \{1 + O(n^{-\delta})\} \quad \text{(by (12.4))} \\ &= 2\sigma_p\, (a_n/2)^{n+1/p}\, G[W(a_n t)]\, \{1 + O(n^{-\delta_1})\}, \end{aligned}$$

some $\delta_1 > 0$, by Lemma 3.3(f), since

$$c_n/a_n = 1 + o\{(\log n)/n\}^{2/3} = 1 + n^{-1/29}\, o(\{nT(a_{2n})\}^{-1/2}),$$

by (3.25). Hence the asymptotic upper bound (12.2) is valid for all $0 < p \le \infty$. For the corresponding lower bound needed for (12.1), we have

$$\begin{aligned} E_{np}(W) &\ge \inf_{P \in \mathcal{P}_{n-1}} \| \{t^n - P(t)\} W(t) \|_{L_p[-c_n,c_n]} \\ &= c_n^{n+1/p} \inf_{P \in \mathcal{P}_{n-1}} \| \{t^n - P(t)\} W(c_n t) \|_{L_p[-1,1]}. \end{aligned}$$

Using the polynomials $S_n^-(t)$ of Theorem 10.1 with $\kappa = 1 - 1/p$, $1 \le p \le \infty$, and (11.3) leads to the corresponding lower bound as before. □

Now we turn to the asymptotics of the L_p extremal polynomials for $2 \le p \le \infty$.

The case $1 < p < 2$ will be handled in Section 14.

Theorem 12.2

Let $W \in SE(3)$ and let $2 \leq p \leq \infty$. Let

$$F_n(x) := W(a_n x)\,(1 - x^2)^{1/(2p)},\ x \in (-1,1). \tag{12.5}$$

Then there exists $\delta > 0$ such that uniformly in closed subsets of $\mathbb{C}\backslash[-1,1]$, as $n \to \infty$,

$$T_{np}(W,a_n z)/\{[a_n\varphi(z)/2]^n\, G[F_n(x)]\, D^{-2}(F_n(\cos t);\tfrac{1}{\varphi(z)})\} = 1 + O(n^{-\delta}), \tag{12.6}$$

and

$$p_{np}(W,a_n z)/\{a_n^{-1/p}\, \varphi(z)^n\, D^{-2}(F_n(\cos t);\tfrac{1}{\varphi(z)})\} = (2\sigma_p)^{-1}\{1 + O(n^{-\delta})\}. \tag{12.7}$$

The value of δ is independent of the particular closed subset.

Proof

We use Lemma 11.2 and the polynomials of Theorem 10.1. Let s be some fixed positive number, let

$$c_n := a_n(1 + s\Delta_n),\ n \geq 1. \tag{12.8}$$

By Theorem 5.2, if s is large enough,

$$\|PW\|_{L_p(\mathbb{R})} \leq (1 + n^{-2})\, \|PW\|_{L_p[-c_n,c_n]},\ n \geq n_1,\ P \in \mathcal{P}_n,\ 2 \leq p \leq \infty. \tag{12.9}$$

According to Theorem 10.1, we can find polynomials $S_n^-(t)$ of degree at most $2n - 2n^{1/4} - 2$, positive in $[-1,1]$, except possibly for simple zeros at ± 1, satisfying (12.3) and (12.4). Let $V_{p,n}$ be as at (12.3), and let in Lemma 11.2,

$$P(z) := c_n^{-1/p}\, p_{np}(V_{p,n},z/c_n)\ (= c_n^{-1/p}\, T_{np}(V_{p,n},z/c_n)/E_{np}(V_{p,n})\ ),$$

a polynomial of degree n, with leading coefficient

$$A := c_n^{-n-1/p}\, E_{np}(V_{p,n})^{-1}$$

$$= (2\sigma_p)^{-1}\, (c_n/2)^{-n-1/p}\, G[V_{p,n}]^{-1} \quad \text{(by (11.3))}$$

$$= (2\sigma_p)^{-1} (a_n/2)^{-n-1/p} G[W(a_n x)]^{-1} \{1 + O(n^{-\delta})\}$$

$$= E_{np}(W)^{-1} \{1 + O(n^{-\delta})\}, \ n \to \infty, \tag{12.10}$$

some $\delta > 0$, exactly as in the proof of Theorem 12.1. Thus

$$r_n := AE_{np}(W) = 1 + O(n^{-\delta}), \ n \to \infty. \tag{12.11}$$

Next, by (12.9),

$$\|PW\|_{L_p(\mathbb{R})} \le (1 + n^{-2}) \|PW\|_{L_p[-c_n,c_n]}$$

$$= (1 + n^{-2}) \|p_{np}(V_{p,n},t)W(c_n t)\|_{L_p[-1,1]}$$

$$\le (1 + n^{-2}) \|p_{np}(V_{p,n},t)V_{p,n}(t)\|_{L_p[-1,1]} = 1 + n^{-2},$$

by (12.3) and by the normalisation of p_{np}. Then (11.12) shows that for $z \in \mathbb{C}\backslash\{x_{jn}^{(p)}\}_{j=1}^n$,

$$|c_n^{-1/p} p_{np}(V_{p,n},z/c_n)/p_{np}(W,z) - 1|$$

$$\le 2\{(1 + n^{-2})^2 - (1 - C_1 n^{-\delta})^2\}^{1/2} / d_n(z) + C_1 n^{-\delta}$$

$$\le C_2 n^{-\delta/2}/d_n(z). \tag{12.12}$$

Here, from (11.1), (11.2), (11.3) and (11.8), for $u \in \mathbb{C}\backslash[-1,1]$,

$$|p_{np}(V_{p,n},u)/\{(2\sigma_p)^{-1} \varphi(u)^n D^{-2}(V_{p,n}(\cos t)|\sin t|^{1/p};\tfrac{1}{\varphi(u)})\} - 1|$$

$$\le |\varphi(u)|^{-2<n^{1/4}>}. \tag{12.13}$$

Now

$$D^{-2}(V_{p,n}(\cos t)|\sin t|^{1/p};\tfrac{1}{\varphi(u)})/D^{-2}(W(c_n\cos t)|\sin t|^{1/p};\tfrac{1}{\varphi(u)})$$

$$= \exp\left(\frac{1}{2\pi}\int_{-\pi}^{\pi} \log[W(c_n\cos t)/V_{p,n}(\cos t)] \frac{1 + \varphi(u)^{-1}e^{-it}}{1 - \varphi(u)^{-1}e^{-it}} dt\right)$$

$$= 1 + O(n^{-\delta}),$$

uniformly for u in closed subsets of $\mathbb{C}\backslash[-1,1]$ (by (12.4)). In view of Theorem 7.1 (a), (12.12) and (12.13) yield

$$p_{np}(W,c_nz)$$
$$= c_n^{-1/p}(2\sigma_p)^{-1}\ \varphi(z)^n\ D^{-2}(W(c_n\cos t)|\sin t|^{1/p};\tfrac{1}{\varphi(z)})\ \{1 + O(n^{-\delta})\},$$
(12.14)

$n \to \infty$, uniformly in closed subsets of $\mathbb{C}\backslash[-1,1]$. The δ does not depend on the closed subset, but the constant in the order term does – note that

$$|\varphi(u)|^{-2<n^{1/4}>} \le \exp(-C\ n^{1/4}),$$

uniformly for u in a closed subset of $\mathbb{C}\backslash[-1,1]$, and n large enough. This establishes (12.7), but with a_n there replaced by c_n. To pass from c_n to a_n, we adopt an indirect method using $p = \infty$, and the sharp L_∞ infinite–finite range identity (5.1). (Direct methods are messier). A glance at the above proof indicates that the only place where we really needed $s > 0$ in the definition (12.8) of c_n, was in estimating $\|PW\|_{L_p(\mathbb{R})}$, and in satisfying (11.9) (the infinite–finite range inequality) of Lemma 11.2. Let us keep c_n as above, so that Lemma 11.2 is applicable, but choose from Theorem 10.1, polynomials $S_n^-(t)$ of degree at most $2n - 2n^{1/4} - 2$, such that

$$W^2(a_nt) \le (1 - t^2)/S_n^-(t) =: V^2_{\infty,n}(t),\ t \in [-1,1],$$

and

$$\int_{-1}^{1} |\log (W^2(a_nt)/V^2_{\infty,n}(t))|\ (1 - t^2)^{-1/2}\ dt = O(n^{-\delta}).$$

In Lemma 11.2, choose

$$P(z) := p_{n\infty}(V_{n\infty},z/a_n)\ (= T_{n\infty}(V_{\infty,n},z/a_n)/E_{n\infty}(V_{\infty,n})),$$

a polynomial of degree n, with leading coefficient

$$A := a_n^{-n}E_{n\infty}(V_{\infty,n})^{-1} = E_{n\infty}(W)^{-1}\ \{1 + O(n^{-\delta})\},\ n \to \infty.$$

– see (12.1) and the proof thereof. Further by (5.1),

$$\|PW\|_{L_\infty(\mathbb{R})} = \|PW\|_{L_\infty[-a_n,a_n]}$$
$$= \|p_{n\infty}(V_{\infty,n},t)W(a_nt)\|_{L_\infty[-1,1]}$$

$$\leq \|p_{n\infty}(V_{\infty,n},t)V_{\infty,n}(t)\|_{L_\infty[-1,1]} = 1.$$

Then, exactly as before, Lemma 11.2 yields, uniformly in closed subsets of $\mathbb{C}\backslash[-1,1]$,

$$p_{n\infty}(W,a_n z) = (2\sigma_\infty)^{-1} \varphi(z)^n D^{-2}(W(a_n \cos t);\frac{1}{\varphi(z)}) \{1 + O(n^{-\delta})\}.$$

(In the conclusion of Lemma 11.2, the factor c_n plays a role only in $d_n(z)$). On the other hand, replacing z by $a_n z/c_n$ in (12.14) for $p = \infty$ leads to

$$p_{n\infty}(W,a_n z) = (2\sigma_\infty)^{-1} \varphi(a_n z/c_n)^n D^{-2}(W(c_n \cos t);\frac{1}{\varphi(a_n z/c_n)}) \{1 + O(n^{-\delta})\}.$$

Hence

$$\frac{\varphi(z)^n D^{-2}(W(a_n \cos t);1/\varphi(z))}{\varphi(a_n z/c_n)^n D^{-2}(W(c_n \cos t);1/\varphi(a_n z/c_n))} = 1 + O(n^{-\delta}), \tag{12.15}$$

as $n \to \infty$, uniformly in closed subsets of $\mathbb{C}\backslash[-1,1]$. On replacing z by $a_n z/c_n$ in (12.14), using (12.15), and noting that uniformly in closed subsets of $\mathbb{C}\backslash[-1,1]$,

$$D^{-2}(|\sin t|^{1/p};1/\varphi(z))/D^{-2}(|\sin t|^{1/p}; 1/\varphi(a_n z/c_n)) = 1 + O(n^{-\delta}),$$

for some small enough δ, we obtain (12.7). Of course, (12.6) then follows, using Theorem 12.1 and the identity $G[1 - x^2] = 1/4$. □

13 Asymptotics of Orthogonal Polynomials on IR

First, a general result about asymptotics of o.p.'s for weights like $G_{Q/2}^{-2}$:

__Theorem 13.1__

Let $W \in SE$, and let $\{c_n\}_{n=1}^{\infty}$ and $\{k_n\}_{n=1}^{\infty}$ be as in Theorem 9.1. Let $G_{Q/2}$ be given by (1.15). For n = 1, 2, 3, ..., let $S_n(x)$ be a polynomial positive in [–1,1] and of degree at most $2k_n$. Let $0 < \beta < 2$ and for $n \geq 1$, let

$$f_n(\theta) := |\sin\theta|^{\beta}/\{G_{Q/2}^2(c_n \cos\theta)S_n(\cos\theta)\},\ \theta \in [-\pi,\pi]. \tag{13.1}$$

Let $L > 0$. Then there exist n_1 and $\delta > 0$ such that uniformly for $n \geq n_1$, $\phi \in [n^{-\delta}, \pi - n^{-\delta}]$, and

$$m \geq 2n - 2Ln\Delta_n, \tag{13.2}$$

we have

$$|\phi_m(f_n,z)/\{z^m/\overline{D(f_n;z)}\} - 1| \leq n^{-\delta},\ z := e^{i\phi}. \tag{13.3}$$

__Proof__

We shall use Lemma 11.6 and the polynomials of Theorem 9.1, with $\kappa = -\beta/2$ and $\kappa = (2-\beta)/2$. By Theorem 9.1, there exist for n large enough, polynomials $P_n^+(u)$, $P_n^-(u)$ of degree at most $2n - 2k_n - 2Ln\Delta_n - 2$, which are positive in [–1,1] (note that we use (9.4) with $\kappa < 0$), and satisfy

$$(1-u^2)^{(2-\beta)/2}/P_n^-(u) \leq 1/G_{Q/2}^2(c_n u) \leq (1-u^2)^{-\beta/2}/P_n^+(u),\ u \in (-1,1).$$

Further, (9.6) to (9.9) are valid. Then defining

$$f_n^+(\theta) := 1/\{P_n^+(\cos\theta)S_n(\cos\theta)\},\ \theta \in [-\pi,\pi],$$

and

$$f_n^-(\theta) := (\sin\theta)^2/\{P_n^-(\cos\theta)S_n(\cos\theta)\},\ \theta \in [-\pi,\pi],$$

we have from (13.1),

$$f_n^-(\theta) \le f_n(\theta) \le f_n^+(\theta),\ \theta \in [-\pi,\pi]. \tag{13.4}$$

The degree of the trigonometric polynomial $P_n^{\pm}(\cos\theta)S_n(\cos\theta)$ is at most $2n - 2Ln\Delta_n - 2$. Next,

$$D(f_n^+;0)/D(f_n^-;0) = \exp\left(\frac{1}{2\pi}\int_0^{\pi} \log\,[f_n^+(\theta)/f_n^-(\theta)]\,d\theta \right)$$

$$= \exp\left(\frac{1}{2\pi}\int_0^{\pi} \log\,[f_n^+(\theta)/f_n(\theta)]\,d\theta + \frac{1}{2\pi}\int_0^{\pi} \log\,[f_n(\theta)/f_n^-(\theta)]\,d\theta\right).$$

Here, with an obvious modification of the notation in Theorem 9.1,

$$f_n(\theta)/f_n^+(\theta) = \{P_n^+(\cos\theta)/G_{Q/2}^2(c_n\cos\theta)\}\,|\sin\theta|^{\beta} =: \tau_n^+(\cos\theta),$$

and

$$f_n(\theta)/f_n^-(\theta) = \{P_n^-(\cos\theta)/G_{Q/2}^2(c_n\cos\theta)\}\,|\sin\theta|^{-(2-\beta)} =: \tau_n^-(\cos\theta).$$

According to (9.8) for τ_n^+ and τ_n^-, we then have

$$D(f_n^+;0)/D(f_n^-;0) = \exp(\,O(n^{-\delta})\,) = 1 + O(n^{-\delta}),\ n \to \infty. \tag{13.5}$$

Next, from (9.6), given $x_0 = \cos\phi$, where for some $\delta_1 > 0$, $\phi \in [n^{-\delta_1}, \pi - n^{-\delta_1}]$,

$$f_n^-(\phi) = f_n(\phi) = f_n^+(\phi). \tag{13.6}$$

Further,

$$I := \int_{-\pi}^{\pi} \frac{\log\{f_n^+(t)/f_n^-(t)\}}{(\sin(\frac{t-\phi}{2}))^2}\,dt$$

$$= \int_0^{\pi} \log\{\tau_n^-(\cos t)/\tau_n^+(\cos t)\}\,\{[\sin(\tfrac{t-\phi}{2})]^{-2} + [\sin(\tfrac{t+\phi}{2})]^{-2}\}\,dt$$

$$\le \pi^2 \int_0^{\pi} \frac{\log\{\tau_n^-(\cos t)/\tau_n^+(\cos t)\}}{(t-\phi)^2}\,dt$$

$$+ C_1 n^{2\delta_1} \int_0^{\pi} \log\{\tau_n^-(\cos t)/\tau_n^+(\cos t)\}\,dt$$

(as $|\sin(\frac{t-\phi}{2})| \ge \pi^{-1}\,|t-\phi|$ and $|\sin(\frac{t+\phi}{2})| \ge C_2 n^{-\delta_1}$)

$$\le \pi^2 n^{\delta/2} \int_{|t-\phi|\ge n^{-\delta/4}} \log\{\tau_n^-(\cos t)/\tau_n^+(\cos t)\}\, dt$$

$$+ \pi^2 \int_{|t-\phi|\le n^{-\delta/4}} \tfrac{1}{2}\left|\left(\tfrac{d}{dt}\right)^2[\log \tau_n^-(\cos t) - \log \tau_n^+(\cos t)]_{|t=\xi}\right| dt$$

$$+ C_1 n^{2\delta_1} \int \log\{\tau_n^-(\cos t)/\tau_n^+(\cos t)\}\, dt.$$

In the second integral, we have expanded $\log \tau_n^-(\cos t) - \log \tau_n^+(\cos t)$ about $t = \phi$ in a Taylor series to second derivatives, and have used (9.6). The number $\xi = \xi(n,t,\phi)$ lies between t and ϕ. Using (9.7) and (9.8), we obtain

$$I \le 50\, n^{-\delta/2} + C\, n^{-\delta/4-\delta} + C\, n^{2\delta_1-\delta}.$$

Then Lemma 11.6 shows that if δ_1 is small enough, then uniformly for $n \ge n_1$, $\phi \in [n^{-\delta_1}, \pi - n^{-\delta_1}]$, $z := e^{i\phi}$ and m satisfying (13.2),

$$|\phi_m(f_n,z)/\{z^m/\overline{D(f_n^+;z)}\} - 1| \le C\, n^{-\delta/2}. \tag{13.7}$$

It remains to show that for $n \ge n_1$ and uniformly for the relevant range of z,

$$|D(f_n^+;z)/D(f_n;z) - 1| = |D(f_n^+/f_n;z) - 1| \le n^{-\delta}. \tag{13.8}$$

According to (11.24) in Lemma 11.3(b), since (13.6) holds, this is equivalent to estimation of

$$J := \int_{-1}^{1} \left|\frac{\log \tau_n^+(s) - \log \tau_n^+(x)}{s - x}\right| (1-s^2)^{-1/2}\, ds, \quad x = \cos \phi.$$

This may be estimated using (9.7) and (9.8) in a somewhat easier fashion than I, and leads to (13.8). □

Theorem 13.2

Let $W \in SE(3)$ and let Q' be continuous at 0. Assume that $\{S_n\}_{n=1}^{\infty}$, $\{c_n\}_{n=1}^{\infty}$, $\{k_n\}_{n=1}^{\infty}$, $G_{Q/2}$, β, and L are as in Theorem 13.1. Let

$$g_n(\theta) := |\sin \theta|^{\beta}\, W^2(c_n \cos \theta)/S_n(\cos\theta), \quad \theta \in [-\pi,\pi]. \tag{13.9}$$

Then there exist n_1 and $\delta > 0$ such that uniformly for $n \geq n_1$, $\phi \in [n^{-\delta}, \pi - n^{-\delta}]$, and m satisfying (13.2),

$$|\phi_m(g_n, z)/\{z^m/\overline{D(g_n; z)}\} - 1| \leq n^{-\delta}, \; z := e^{i\phi}. \tag{13.10}$$

<u>Proof</u>

We shall use Theorem 13.1, Lemma 10.2 and Lemma 11.8. Let $\{R_n\}_{n=1}^{\infty}$ be the polynomials of degree $\leq n^{3/29}$ of Lemma 10.2. We apply Theorem 13.1 with

$$f_n(\theta) := |\sin \theta|^{\beta}/\{G_{Q/2}^2(c_n \cos \theta)\, S_n(\cos \theta)\, R_n^2(\cos \theta)\}.$$

As $S_n R_n^2$ has degree $O(n\Delta_n)$, the degree requirements of Theorem 13.1 are met. Then (13.3) implies that for the range (13.2) of m,

$$|\phi_m(f_n, e^{i\phi})\, f_n(\phi)^{1/2}| \leq 2, \; \phi \in [n^{-\delta}, \pi - n^{-\delta}], \; n \geq n_1.$$

Let us take in Lemma 11.8, $[a,b] = [n^{-\delta}, \pi - n^{-\delta}]$, so that in (11.66) we may take M = 2. Also, uniformly for $\theta \in [-\pi, \pi]$, with the notation (10.9),

$$\begin{aligned} f_n(\theta)/g_n(\theta) &= R_n^{-2}(\cos \theta)\, \{G_{Q/2}W\}^{-2}(c_n \cos \theta) \\ &= \{\Psi(c_n \cos \theta)/R_n(\cos \theta)\}^2 \\ &= 1 + O(n^{-\delta_1}), \end{aligned} \tag{13.11}$$

for some fixed $\delta_1 > 0$, (by (10.12) in Lemma 10.2). Further by (10.13),

$$(f_n/g_n)'(\theta) = O(n^{16/29}), \; \theta \in [-\pi, \pi]. \tag{13.12}$$

Thus, in Lemma 11.8, we may take

$$\eta = O(n^{-\delta_1}); \; \nu = O(n^{16/29}).$$

Further, if for some small enough $\delta_2 > 0$, $\phi \in (n^{-\delta_2}, \pi - n^{-\delta_2})$, then β of (11.67) satisfies $\beta \geq Cn^{-\delta_2}$, and hence B of (11.70) satisfies

$$B = O(n^{-\delta_1 + \delta_2} \log n) = O(n^{-\delta_1/2}).$$

Finally in Lemma 11.8, let

$$\epsilon := n^{-\delta_1/4}.$$

Choosing j fixed, but so large that

$$j - 1 > 32/(29\delta_1),$$

Lemma 11.8 ensures that for $\phi \in [n^{-\delta}+jn^{-\delta_1/2}, \pi-n^{-\delta}-jn^{-\delta_1/2}]$,

$$|\phi_m(g_n,e^{i\phi})/\phi_m(f_n,e^{i\phi}) - 1|$$

$$= O(n^{-\delta_1+\delta_1/4+\delta_2}) + O(n^{-\delta_1/2}) + O(n^{\delta_2+8/29-\delta_1/2-(j-1)\delta_1/2})$$

$$= O(n^{-\delta_1/2}).$$

In order to complete the proof of the theorem, it suffices to apply (13.3), and then to show that $D(f_n;z)/D(g_n;z)$ is close to 1. But by (11.24), in Lemma 11.3(b), after making there the substitutions $s = \cos\theta$; $x = \cos\phi$,

$$|D(f_n;z)/D(g_n;z) - 1| = |D(f_n/g_n;z) - 1|$$

$$\leq |\log(f_n/g_n)(\phi)| + \frac{\sin\phi}{\pi}\int_0^{\pi}\left|\frac{\log(f_n/g_n)(\theta)-\log(f_n/g_n)(\phi)}{\cos\theta-\cos\phi}\right| d\theta$$

$$= O(n^{-\delta_1}) + O(n^{-\delta_1}\log n),$$

uniformly for $\phi \in (0,\pi)$, by (13.11) and (13.12). □

We must turn the asymptotics on the circle into asymptotics on the real line, but first must compare orthogonal polynomials for weights on $\mathbb{R}$ and their truncations:

Lemma 13.3

Let $W \in SE$, let $K > 0$, and let $\{c_n\}_{n=1}^{\infty}$ be given by (9.1), with s large enough. Let

$$\Omega_n(x) := \begin{cases} W^2(c_nx), & x \in [-1,1], \\ 0, & \text{otherwise.} \end{cases} \tag{13.13}$$

Then for $n \geq n_1$ and for $1 \leq m \leq n$,

$$\|c_n^{1/2}p_m(W^2,c_nx)W(c_nx) - p_m(\Omega_n,x)W(c_nx)\|_{L_\infty(\mathbb{R})} \leq n^{-K}. \tag{13.14}$$

Proof

With standard notation, let

$$K_m(\Omega_n,x,t) := \sum_{j=0}^{m-1} p_j(\Omega_n,x)\, p_j(\Omega_n,t)$$

$$= \{\gamma_{m-1}(\Omega_n)/\gamma_m(\Omega_n)\} \frac{p_m(\Omega_n,x)p_{m-1}(\Omega_n,t) - p_m(\Omega_n,t)p_{m-1}(\Omega_n,x)}{x - t}.$$

Then, using orthogonality, we see that

$$\pi_{m-1}(x) := c_n^{1/2} p_m(W^2,c_n x) - c_n^{m+1/2} p_m(\Omega_n,x)\gamma_m(W^2)/\gamma_m(\Omega_n) \tag{13.15}$$

$$= \int_{-1}^{1} K_m(\Omega_n,x,t)\, \{c_n^{1/2} p_m(W^2,c_n t)\}\, \Omega_n(t)\, dt$$

$$= \int_{-c_n}^{c_n} K_m(\Omega_n,x,u/c_n)\, p_m(W^2,u)\, W^2(u)\, du\, c_n^{-1/2}$$

$$= -\{\gamma_{m-1}(\Omega_n)/\gamma_m(\Omega_n)\}\, c_n^{-1/2}$$

$$\times \{p_m(\Omega_n,x) \int_{|u|>c_n} \frac{p_{m-1}(\Omega_n,u/c_n)p_m(W^2,u)}{x - u/c_n} W^2(u)\, du$$

$$- p_{m-1}(\Omega_n,x) \int_{|u|>c_n} \frac{p_m(\Omega_n,u/c_n)p_m(W^2,u)}{x - u/c_n} W^2(u)\, du\}.$$

Since Ω_n is supported on $[-1,1]$, [4,p.41,proof of Lemma 7.2],

$$\gamma_{m-1}(\Omega_n)/\gamma_m(\Omega_n) \le 1.$$

Then if s in (9.1) is large enough, and $|x| < 1$, and $m \le n$, we obtain from Theorem 5.2,

$$|\pi_{m-1}(x)W(c_n x)| \le c_n^{-1/2} (1 - |x|)^{-1} \{|p_m(\Omega_n,x)W(c_n x)|$$

$$\times \|p_{m-1}(\Omega_n,u/c_n)W(u)\|_{L_2(|u|>c_n)} \|p_m(W^2,u)W(u)\|_{L_2(\mathbb{R})}$$

$$+ |p_{m-1}(\Omega_n,x)W(c_n x)|\, \|p_m(\Omega_n,u/c_n)W(u)\|_{L_2(|u|>c_n)}$$

$$\times \|p_m(W^2,u)W(u)\|_{L_2(\mathbb{R})}\}$$

$$\leq c_n^{-1/2} (1 - |x|)^{-1} n^{-2K}$$

$$\times \{|p_m(\Omega_n,x)W(c_nx)| \, \|p_{m-1}(\Omega_n,u/c_n)W(u)\|_{L_2(|u|\leq c_n)}$$

$$+ |p_{m-1}(\Omega_n,x)W(c_nx)| \, \|p_m(\Omega_n,u/c_n)W(u)\|_{L_2(|u|\leq c_n)}\}$$

$$= (1 - |x|)^{-1} n^{-2K} \{|p_m(\Omega_n,x)W(c_nx)| \, \|p_{m-1}(\Omega_n,t)\Omega_n^{1/2}(t)\|_{L_2[-1,1]}$$

$$+ |p_{m-1}(\Omega_n,x)W(c_nx)| \, \|p_m(\Omega_n,t)\Omega_n^{1/2}(t)\|_{L_2[-1,1]}\}$$

$$\leq (1 - |x|)^{-1} n^{-2K} \sum_{j=m-1}^{m} |p_j(\Omega_n,x)W(c_nx)| . \tag{13.16}$$

Next, by a crude Nikolskii inequality [14,Lemma 7.3,p.53] and then by Theorem 5.2,

$$\|p_j(\Omega_n,x/c_n)W(x)\|_{L_\infty(\mathbb{R})} \leq C\, n^2 \, \|p_j(\Omega_n,x/c_n)W(x)\|_{L_2(\mathbb{R})}$$

$$\leq 2C\, n^2 \, \|p_j(\Omega_n,x/c_n)W(x)\|_{L_2[-c_n,c_n]}$$

$$= 2C\, n^2\, c_n^{1/2} \, \|p_j(\Omega_n,x)\Omega_n^{1/2}(x)\|_{L_2[-1,1]}$$

$$= 2C\, n^2\, c_n^{1/2}.$$

Thus for $n \geq n_1$,

$$\|p_j(\Omega_n,x/c_n)W(x)\|_{L_\infty(\mathbb{R})} \leq n^3, \; j = m-1, m; \; 1 \leq m \leq n. \tag{13.17}$$

Then from (13.16), for $|x| \leq a_n/c_n = (1 + s\Delta_n)^{-1}$,

$$|\pi_{m-1}(x)W(c_nx)| \leq 2(s\Delta_n)^{-1} n^{-2K+3} \leq n^{-2K+4},$$

$n \geq n_1$. Making the substitution $c_nx = a_nu$, using (5.1), and then resubstituting yields

$$|\pi_{m-1}(x)W(c_nx)| \leq n^{4-2K}, \; x \in \mathbb{R}. \tag{13.18}$$

Next, from Theorem 5.2, if s is large enough, and $n \geq n_1$, and $1 \leq m \leq n$,

$$\gamma_m(W^2)^{-1} = \inf_{P\in \mathcal{P}_{m-1}} \|\{x^m - P(x)\}W(x)\|_{L_2(\mathbb{R})}$$

$$\leq (1 + n^{-2K}) \inf_{P\in \mathcal{P}_{m-1}} \|\{x^m - P(x)\}W(x)\|_{L_2[-c_n,c_n]}$$

$$= (1 + n^{-2K})\, c_n^{m+1/2} \inf_{P\in \mathcal{P}_{m-1}} \|\{t^m - P(t)\}W(c_n t)\|_{L_2[-1,1]}$$

$$= (1 + n^{-2K})\, c_n^{m+1/2}\, \gamma_m(\Omega_n)^{-1}.$$

Similarly

$$\gamma_m(W^2)^{-1} \geq c_n^{m+1/2}\, \gamma_m(\Omega_n)^{-1},$$

so

$$|\, \gamma_m(W^2)\, c_n^{m+1/2}/\gamma_m(\Omega_n) - 1| \leq n^{-2K}.$$

Hence from (13.15) and (13.18), for $x \in \mathbb{R}$, and $1 \leq m \leq n$,

$$|\, c_n^{1/2}\, p_m(W^2,c_n x)W(c_n x) - p_m(\Omega_n,x)W(c_n x)|$$

$$\leq n^{-2K+4} + n^{-2K}\, |\, p_m(\Omega_n,x)W(c_n x)|$$

$$\leq n^{-2K+4} + n^{-2K+3},$$

by (13.17). Choosing K large enough, we obtain (13.14). □

We note that an alternative (and shorter) proof of Lemma 13.3 may be based on the parallelogram law. Theorem 2.4 is the special case $m = n$ of:

<u>Theorem 13.4</u>

Let $W \in SE(3)$, and let Q' be continuous at 0. Let $L > 0$, and let $\{c_n\}_{n=1}^{\infty}$ be given by (9.1), with s large enough. Let $\{\hat{F}_n(\cdot)\}_{n=1}^{\infty}$ be defined by (2.12) with $p = 2$. Then there exist n_1 and $\delta > 0$ s.t. for $n \geq n_1$,

$$c_n^{1/2}\, p_m(W^2,c_n x)\, \hat{F}_n(x)$$

$$= (\tfrac{2}{\pi})^{1/2}\, \mathrm{Re}\, [z^m\, \hat{F}_n(\cos\phi)/\overline{D(\hat{F}_n^2(\cos\ t);z)}] + O(n^{-\delta}), \tag{13.19}$$

uniformly for

$$x = \cos\phi,\ \phi \in [n^{-\delta}, \pi - n^{-\delta}];\quad z := e^{i\phi}, \tag{13.20}$$

and

$$n \geq m \geq n - \langle Ln\Delta_n \rangle. \tag{13.21}$$

We may also write

$$c_n^{1/2}\, p_m(W^2, c_n x)\, \hat{F}_n(x) = (\tfrac{2}{\pi})^{1/2}\cos(m\phi + \Gamma(\hat{F}_n^2;\phi)) + O(n^{-\delta}), \tag{13.22}$$

$$= (\tfrac{2}{\pi})^{1/2}\cos((m-n)\phi + n\pi\int_x^1 \mu_{n,c_n}(t)\,dt + \tfrac{\phi}{2} - \tfrac{\pi}{4}) + O(n^{-\delta}), \tag{13.23}$$

uniformly for x and m as in (13.20) and (13.21). Here μ_{n,c_n} is defined by (4.3) to (4.5).

Proof

Let $\Omega_n(x)$ be defined by (13.13), and let

$$\mu_n'(\theta) := W^2(c_n \cos\theta)\,|\sin\theta| = \Omega_n(\cos\theta)\,|\sin\theta| = \hat{F}_n^2(\cos\theta),\ \theta \in [-\pi,\pi].$$

We have [30,p.294,eqn.(11.5.2)],

$$p_m(\Omega_n, x) = (2\pi)^{-1/2}\,\{1 + \phi_{2m}(\mu_n',0)/\kappa_{2m}(\mu_n')\}^{-1/2}$$

$$\times\ \{z^{-m}\,\phi_{2m}(\mu_n',z) + z^m\,\phi_{2m}(\mu_n',z^{-1})\}$$

$$= (2\pi)^{-1/2}\,\{1 + \phi_{2m}(\mu_n',0)/\kappa_{2m}(\mu_n')\}^{-1/2}$$

$$\times\ \{z^m/\overline{D(\mu_n';z)} + z^{-m}/\overline{D(\mu_n';1/z)}\}\,\{1 + O(n^{-\delta})\},$$

uniformly for x and m satisfying (13.20–21), by Theorem 13.2, with $\beta = 1$ and $S_n = 1$. Since

$$|D(\mu_n';1/z)| = |D(\mu_n';z)| = \mu_n'(\phi)^{1/2} = \hat{F}_n(x),$$

we have, again uniformly for x and m as above,

$$p_m(\Omega_n, x)\,\hat{F}_n(x) = (2\pi)^{-1/2}\,\{1+\phi_{2m}(\mu_n',0)/\kappa_{2m}(\mu_n')\}^{-1/2}$$

$$\times\ \{z^m\hat{F}_n(x)/\overline{D(\hat{F}_n^2(\cos t);z)} + z^{-m}\hat{F}_n(x)/\overline{D(\hat{F}_n^2(\cos t);\tfrac{1}{z})}$$

$$+ O(n^{-\delta})\}.$$

It remains to show that

$$\Phi_{2m}(\mu_n',0) := \phi_{2m}(\mu_n',0)/\kappa_{2m}(\mu_n')$$

satisfies

$$\max\{\ |\Phi_{2m}(\mu_n',0)| : n - <Ln\Delta_n> \leq m \leq n\} = O(n^{-\delta}),\ n \to \infty, \tag{13.24}$$

and then to apply Lemma 13.3 to complete the proof of (13.19). From (11.16) and (11.46) (with $R = c_n$), we can then deduce (13.22) and (13.23). There are several ways to establish (13.24), the simplest being to use an inequality of Rahmanov [28,p.150,Theorem 1]:

$$|\Phi_k(\mu_n',0)| \leq 2\,[\inf_{P\in \mathcal{P}_k} \frac{1}{2\pi}\int_{-\pi}^{\pi} \{|P(e^{i\theta})|\ \mu_n'(\theta)^{1/2} - 1\}^2\, d\theta]^{1/2}$$

$$\leq 2\,[\inf_{\substack{S\in \mathcal{P}_k \\ S\geq 0}} \frac{1}{2\pi}\int_{-\pi}^{\pi} \{[S(\cos\theta)W^2(c_n\cos\theta)\ |\sin\theta|\,]^{1/2} - 1\}^2 d\theta]^{1/2}. \tag{13.25}$$

Here the second inf is taken over all even trigonometric polynomials $S(\cos\theta)$ of degree at most k, non–negative in $[-1,1]$. The equivalence of $|P(e^{i\theta})|^2$ and $S(\cos\theta)$ is discussed in [30,p.3]. But according to Theorem 10.1, we can find polynomials S_n of degree at most $2n - 2k_n$, provided only $k_n = O(n\Delta_n)$, such that

$$S_n(t)W^2(c_nt)\ (1 - t^2)^{1/2} \leq 1,\ t \in (-1,1),$$

and for some $\delta > 0$,

$$|S_n(t)W^2(c_nt)\ (1 - t^2)^{1/2} - 1| \leq n^{-2\delta},\ |t| \leq 1 - n^{-2\delta}.$$

Applying these to (13.25) yields (13.24) for some $\delta > 0$, and hence the theorem. □

In establishing asymptotics for Christoffel functions, we shall need an analogue of Theorem 13.4 for the weight $1/G_{Q/2}^2$. First, however, we need an infinite–finite range inequality adequate for this section, and for Section 15.

Lemma 13.5

Let $W \in SE$ and let $G_{Q/2}$ be the associated entire function defined by (1.15). Let $0 \le p_0 \le 1$ and $K > 0$. Let $\{c_n\}_{n=1}^{\infty}$ be given by (9.1), with s large enough.

(a) Then there exists n_1 such that for $p_0 \le p \le \infty$, $n \ge n_1$ and $P \in \mathcal{P}_n$,

$$\|P/G_{Q/2}\|_{L_p(|x|\ge c_n)} \le n^{-K} \|P/G_{Q/2}\|_{L_p(|x|\le c_n)}, \tag{13.26}$$

and

$$\|P/G_{Q/2}\|_{L_p(\mathbb{R})} \le (1 + n^{-K}) \|P/G_{Q/2}\|_{L_p(|x|\le c_n)}. \tag{13.27}$$

(b) Let $a \in \mathbb{R}$ and let Ψ be given by (10.9), and let

$$\hat{W}(x) := G_{Q/2}(x)^{-1} \Psi(x)^a, \ x \in \mathbb{R}. \tag{13.28}$$

Then, there exists n_1 such that for $p_0 \le p \le \infty$, $n \ge n_1$ and $P \in \mathcal{P}_n$,

$$\|P\hat{W}\|_{L_p(\mathbb{R})} \le (1 + n^{-K}) \|P\hat{W}\|_{L_p(|x|\le c_{2n})}. \tag{13.29}$$

Proof

(a) Recall first that

$$1/G_{Q/2} = W \Psi,$$

and from Theorem 6.1(a), we have in $\mathbb{R}$,

$$\Psi \sim T^{-1/2}.$$

Note that T is increasing in $(0, \infty)$. Hence from Theorem 5.2, if s in (9.1) is large enough, we have for $n \ge n_1$, $P \in \mathcal{P}_n$, and $p_0 \le p \le \infty$,

$$\|P/G_{Q/2}\|_{L_p(|x|\ge c_n)} \le C_1 T(c_n)^{-1/2} \|PW\|_{L_p(|x|\ge c_n)}$$

$$\le C_1 T(c_n)^{-1/2} n^{-2K} \|PW\|_{L_p(|x|\le c_n)}$$

$$\le C_1 n^{-2K} \|PWT^{-1/2}\|_{L_p(|x|\le c_n)}$$

$$\le C_2 n^{-2K} \|P/G_{Q/2}\|_{L_p(|x|\le c_n)}.$$

Hence (13.26). Then (13.27) follows via the triangle inequality, at least if $p_0 = 1$. If

$p_0 < 1$, the range $p_0 \le p \le 1$ may be handled in a slightly more complicated manner.

(b) We can write

$$\hat{W}(x) = W(x)\ \Psi(x)^{a+1},\ x \in \mathbb{R},$$
$$= \exp(-\hat{Q}(x)),\ x \in \mathbb{R},$$

where

$$\hat{Q}(x) := Q(x) - (a+1)\log \Psi(x),\ x \in \mathbb{R}.$$

Let $h(x) := |\log \Psi(x)|/Q(x)$, $x \in (0, \infty)$. Now from (3.25) and (3.35),

$$h(a_u) = O(\frac{\log T(a_u)}{Q(a_u)}) = o(1),\ u \to \infty.$$

For n large enough, let

$$\eta_n := |a+1|\ \max\{\ h(x) : |x| \ge c_{2n}\}.$$

We have $\eta_n \to 0$ as $n \to \infty$, and

$$1 - \eta_n \le \hat{Q}(x)/Q(x) \le 1 + \eta_n,\ |x| \ge c_{2n},$$

and hence

$$W(x)^{1-\eta_n} \ge \hat{W}(x) \ge W(x)^{1+\eta_n},\ |x| \ge c_{2n}. \tag{13.30}$$

Concerning Mhaskar–Rahmanov–Saff numbers for W and $W^{1/2}$, we note that (with an obvious notation), $a_n(W^{1/2}) = a_{2n}(W)$. Further, it follows from (1.12) that

$$\Delta_n(W^{1/2})/\Delta_n(W) = 2^{2/3},\ n \ge 2.$$

Hence, if s in (9.1) is large enough, Theorem 5.2 applied to $W^{1/2}$ yields for $p_0 \le p \le \infty$, $n \ge n_1$ and $P \in \mathcal{P}_n$,

$$\|PW^{1/2}\|_{L_p(|x|\ge c_{2n})} \le n^{-2K}\ \|PW^{1/2}\|_{L_p(|x|\le c_{2n})}$$

$$\le n^{-2K}\ \|P\hat{W}\|_{L_p(|x|\le c_{2n})}\ \|W^{1/2}/\hat{W}\|_{L_\infty(|x|\le c_{2n})}. \tag{13.31}$$

Then from (13.30),

$$\|P\hat{W}\|_{L_p(|x|\ge c_{2n})} \le \|PW^{1-\eta_n}\|_{L_p(|x|\ge c_{2n})}$$

$$\le \|PW^{1/2}\|_{L_p(|x|\ge c_{2n})} \|W^{1/2-\eta_n}\|_{L_\infty(|x|\ge c_{2n})}$$

$$\le n^{-2K} \|P\hat{W}\|_{L_p(|x|\le c_{2n})} \|W^{1/2}/\hat{W}(x)\|_{L_\infty(|x|\le c_{2n})}$$

$$\times \|W^{1/2-\eta_n}\|_{L_\infty(|x|\ge c_{2n})}$$

(by (13.31))

$$\le n^{-2K} \|P\hat{W}\|_{L_p(|x|\le c_{2n})}$$

$$\times \exp(\tfrac{1}{2} Q(c_{2n}) + |a+1| \|\log \Psi\|_{L_\infty(|x|\le c_{2n})} - \tfrac{1}{2} Q(c_{2n}) + Q(c_{2n}) \eta_n),$$

by the definition of $\hat{Q}$. Here, from (3.25),

$$\|\log \Psi\|_{L_\infty(|x|\le c_{2n})} \le C_1 \log n,$$

with C_1 independent of K. Using (3.25) and (3.35), we also see that

$$\eta_n Q(c_{2n})$$

$$\le C_2\{ \max_{|u|\ge n^2} \log T(a_u)/Q(a_u) + \max_{c_{2n}\le |x|\le a_{n^2}} \log T(x)/Q(x)\}Q(c_{2n})$$

$$\le C_3\{ (\log n^2)\, n^{-2(27/29)}\, Q(c_{2n}) + \log n\}$$

$$\le C_4 \log n.$$

Again C_4 is independent of K. Hence, choosing K large enough, yields

$$\|P\hat{W}\|_{L_p(|x|\ge c_{2n})} \le n^{-K} \|P\hat{W}\|_{L_p(|x|\le c_{2n})},$$

for $p_0 \le p \le \infty$, $n \ge n_1$ and $P \in \mathcal{P}_n$. Then the result follows. □

Following is an analogue for $1/G^2_{Q/2}$ of Lemma 13.3.

<u>Lemma 13.6</u>

Let $W \in SE$, let $K > 0$, and let $\{c_n\}_{n=1}^{\infty}$ be given by (9.1), with s large enough. Let

$$\Omega_n(x) := \begin{cases} 1/G^2_{Q/2}(c_nx), & x \in [-1,1], \\ 0 & , \text{otherwise.} \end{cases} \tag{13.32}$$

Then for $n \geq n_1$ and for $1 \leq m \leq n$,

$$\| c_n^{1/2} p_m(G^{-2}_{Q/2}, c_nx)/G_{Q/2}(c_nx) - p_m(\Omega_n, x)/G_{Q/2}(c_nx) \|_{L_\infty(\mathbb{R})} \leq n^{-K}. \tag{13.33}$$

Proof

This is exactly the same as Lemma 13.3. The only difference is that instead of Theorem 5.2, we use Lemma 13.5(a) as our infinite–finite range inequality. □

We can now state:

Theorem 13.7

Let $W \in SE$. Let $L > 0$, and let $\{c_n\}_{n=1}^\infty$ be given by (9.1), with s large enough. Let

$$\tilde{F}_n(x) := G^{-1}_{Q/2}(c_nx)\,(1 - x^2)^{1/4},\ x \in (-1,1),\ n \geq 1. \tag{13.34}$$

Then there exist n_1 and $\delta > 0$ s.t. for $n \geq n_1$,

$$c_n^{1/2}\ p_m(G^{-2}_{Q/2}, c_nx)\ \tilde{F}_n(x)$$

$$= (\tfrac{2}{\pi})^{1/2} \operatorname{Re}\,[z^m\ \tilde{F}_n(\cos\phi)/\overline{D(\tilde{F}^2_n(\cos t);z)}] + O(n^{-\delta}), \tag{13.35}$$

$$= (\tfrac{2}{\pi})^{1/2}\cos(m\phi + \Gamma(\tilde{F}^2_n;\phi)) + O(n^{-\delta}), \tag{13.36}$$

uniformly for x as in (13.20) and m as in (13.21).

Proof

This is exactly the same as that of Theorem 13.4. We merely substitute Theorem 13.1, Lemma 13.6, and Theorem 9.1, respectively for Theorem 13.2, Lemma 13.3, and Theorem 10.1. □

14 Mean Asymptotics of Extremal Polynomials

In this section, we prove the mean asymptotic Theorem 2.2(c), and use it to prove Theorem 2.2(b) for $1 < p < 2$. These results are restated in Theorems 14.3 and 14.5.

As in [15], the method involves Clarkson's inequalities. We first establish two lemmas involving extremal errors $E_{mp}(w)$ and normalised extremal polynomials $p_{mp}(w,x)$ associated with a weight w on [–1,1]. The notation is that of Lemma 11.1, together with an obvious modification of (1.4) to (1.5).

Lemma 14.1

Let $1 < p < \infty$, and let $w \in L_p[-1,1]$ be a non–negative function. Let P(z) be a polynomial of degree $m \geq 1$, with leading coefficient $\Delta \neq 0$, and let

$$A := \Delta\, E_{mp}(w). \tag{14.1}$$

Then

$$\| \{P(x)/A - p_{mp}(w,x)\} w(x) \|_{L_p[-1,1]} \leq \psi(\|Pw\|_{L_p[-1,1]}/A), \tag{14.2}$$

where if $2 \leq p < \infty$,

$$\psi(x) := 2^{1-1/p} \{x^p - 1\}^{1/p},\ x \in [1, \infty), \tag{14.3}$$

and if $1 < p < 2$,

$$\psi(x) := 2\, \{[(x^p + 1)/2]^{1/(p-1)} - 1\}^{(p-1)/p},\ x \in [1, \infty). \tag{14.4}$$

Proof

This appeared first in [15], but since [15] has not yet appeared in print, we provide the details. Recall Clarkson's inequalities [2,p.3]. If $f, g \in L_p[-1,1]$, then for $1 < p < 2$,

$$\|f+g\|^{p/(p-1)}_{L_p[-1,1]} + \|f-g\|^{p/(p-1)}_{L_p[-1,1]}$$
$$\le 2\{\|f\|^p_{L_p[-1,1]} + \|g\|^p_{L_p[-1,1]}\}, \tag{14.5}$$

and if $2 \le p < \infty$,

$$\|f+g\|^p_{L_p[-1,1]} + \|f-g\|^p_{L_p[-1,1]}$$
$$\le 2^{p-1}\{\|f\|^p_{L_p[-1,1]} + \|g\|^p_{L_p[-1,1]}\}. \tag{14.6}$$

Now suppose $1 < p < 2$, and apply (14.5) with $f := (Pw)/\Delta$ and $g := T_{mp}(w,\cdot)w$. Note that $(f+g)/(2w)$ is a monic polynomial of degree m, so that

$$\|(f+g)/2\|_{L_p[-1,1]} \ge E_{mp}(w).$$

Hence from (14.5),

$$\|\{P(x)/\Delta - T_{mp}(w,x)\}w(x)\|^{p/(p-1)}_{L_p[-1,1]}$$
$$\le 2\{\|Pw\|^p_{L_p[-1,1]}/\Delta^p + E^p_{mp}(w)\}^{1/(p-1)} - (2E_{mp}(w))^{p/(p-1)}.$$

Using (14.1), we obtain after dividing by $E_{mp}(w)^{p/(p-1)}$,

$$\|\{P(x)/A - p_{mp}(w,x)\}w(x)\|^{p/(p-1)}_{L_p[-1,1]}$$
$$\le 2\{\|Pw\|^p_{L_p[-1,1]}/A^p + 1\}^{1/(p-1)} - 2^{p/(p-1)}.$$

Then (14.2) follows on using (14.4). The case $2 \le p < \infty$ is similar. □

Lemma 14.2

Let $1 \le p < \infty$ and let $f: [-\pi,\pi] \to [0,\infty)$ satisfy $\log f \in L_1[-\pi,\pi]$ and $f^{-1} \in L_p[-\pi,\pi]$. Let $0 < \delta < 1$. Then

$$\int_{-\pi}^{\pi} |D^{-2}(f;e^{i\theta}) - 1|^p\, d\theta \le C_1 \{\int_{\mathscr{S}} [f(\theta)^{-p} + 1]\, d\theta$$
$$+ \int_{-\pi}^{\pi} |f(\theta)^{-1} - 1|^p\, d\theta + \delta^p \int_{-\pi}^{\pi} f(\theta)^{-p}\, d\theta\}, \tag{14.7}$$

where $\mathscr{S}$ is a measurable subset of $[-\pi,\pi]$ with

$$\text{meas}(\mathscr{S}) \leq C_2\, \delta^{-1} \int_{-\pi}^{\pi} |\log f(\theta)|\; d\theta. \tag{14.8}$$

Here meas denotes linear Lebesgue measure, and C_1 and C_2 are independent of f and δ.

Proof

This also appeared in [15], but we again provide the details. Two standard representations for Szegö's function are (1.18–9) and (11.16). Another form of these representations is [14,pp.278–9]

$$D(f;e^{i\theta}) = f(\theta)^{1/2} \exp(i\, \widehat{\log f}(\theta)), \tag{14.9}$$

where $\widehat{\quad}$ denotes the operation of forming the conjugate function (this is nonstandard notation). That is,

$$\widehat{\log f}(\theta) := \frac{PV}{4\pi} \int_{-\pi}^{\pi} \log f(t) \cot(\frac{\theta - t}{2})\; dt, \text{ a.e. } \theta \in [-\pi,\pi],$$

where PV denotes Cauchy principal value. A classical theorem of Kolmogorov [5,p.111,p.123] asserts that the conjugate function operator is weak type (1,1). In the present context, this implies that if

$$\mathscr{S} := \{\theta \in [0,2\pi]: |\widehat{\log f}(\theta)| > \delta\},$$

then

$$\delta\, \text{meas}(\mathscr{S}) \leq C_2\, \|\log f\|_{L_1[-\pi,\pi]}.$$

Here C_2 is independent of f and δ. Thus (14.8) is true.

Next, from (14.9),

$$\int_{\mathscr{S}} |D^{-2}(f;e^{i\theta}) - 1|^p\; d\theta \leq \int_{\mathscr{S}} [f(\theta)^{-1} + 1]^p\; d\theta.$$

Also, if $\theta \in [-\pi,\pi]\backslash\mathscr{S}$, and since $\delta < 1$,

$$|\exp(-i\, \widehat{\log f}(\theta)) - 1| \leq 2\, |\widehat{\log f}(\theta)| \leq 2\delta,$$

so from (14.9),

$$\int_{[-\pi,\pi]\setminus\mathscr{S}} |D^{-2}(f;e^{i\theta}) - 1|^p \, d\theta$$

$$= \int_{[-\pi,\pi]\setminus\mathscr{S}} |f(\theta)^{-1} [\exp(-i \widetilde{\log f}(\theta)) - 1] + f(\theta)^{-1} - 1|^p \, d\theta$$

$$\leq 2^p \{(2\delta)^p \int_{[-\pi,\pi]\setminus\mathscr{S}} f(\theta)^{-p} \, d\theta + \int_{[-\pi,\pi]\setminus\mathscr{S}} |f(\theta)^{-1} - 1|^p \, d\theta\}.$$

Then (14.7) follows. □

Theorem 2.2(c) is the special case $m = n$ of

Theorem 14.3

Let $W \in SE(3)$. Let $1 < p < \infty$ and let $C_0 > 0$. Let $\{c_n\}_{n=1}^{\infty}$ be as in (2.11), with s fixed but large enough, and let $\{\hat{F}_n\}_{n=1}^{\infty}$ be as in (2.12). Then for some $\delta > 0$, as $n \to \infty$,

$$\| \{c_n^{1/p} \, p_{mp}(W, c_n x) - \frac{1}{\sigma_p} \operatorname{Re}[z^m D^{-2}(\hat{F}_n(\cos t); \tfrac{1}{z})]\} W(c_n x) \|_{L_p[-1,1]}$$

$$= O(n^{-\delta}), \tag{14.10}$$

uniformly for

$$n \geq m \geq n - C_0 \, n \, \Delta_n. \tag{14.11}$$

Here for $x = \cos\theta \in [-1,1]$, we set $z := e^{i\theta}$.

Proof

This will consist of three steps.

Step 1 Replacing $\mathbb{R}$ by $[-1,1]$.

We assume that s in the definition of c_n is so large that (5.8) in Theorem 5.2 holds with $S = 2$. Let

$$W_n(x) := \begin{cases} W(c_n x), & x \in [-1,1], \\ 0, & \text{otherwise.} \end{cases}$$

We apply Lemma 14.1, with

$$P(x) := c_n^{1/p}\, p_{mp}(W,c_n x) \text{ and } w := W_n.$$

Observe that with the notation (14.1),

$$A := A(m) := c_n^{m+1/p}\, E_{mp}(W)^{-1}\, E_{mp}(W_n).$$

Now by (5.8),

$$E_{mp}(W) = \inf_{S\in P_{m-1}} \|\{x^n - S(x)\}W(x)\|_{L_p(\mathbb{R})}$$

$$= \inf_{S\in P_{m-1}} \|\{x^n - S(x)\}W(x)\|_{L_p[-c_n,c_n]}\, (1 + O(n^{-2}))$$

$$= c_n^{m+1/p} \inf_{S\in P_{m-1}} \|\{t^n - S(t)\}W_n(t)\|_{L_p[-1,1]}\, (1 + O(n^{-2}))$$

$$= c_n^{m+1/p}\, E_{mp}(W_n)\, (1 + O(n^{-2})).$$

So

$$A(m) = 1 + O(n^{-2}),\ n \to \infty. \tag{14.12}$$

Further, a substitution $t = c_n x$ shows that

$$\|Pw\|_{L_p[-1,1]} = \|c_n^{1/p}\, p_{mp}(W,c_n x)\, W(c_n x)\|_{L_p[-1,1]}$$

$$= \|p_{mp}(W,t)\, W(t)\|_{L_p[-c_n,c_n]} \le 1.$$

Then, using the above notation, we have

$$\|\{c_n^{1/p}\, p_{mp}(W,c_n x) - p_{mp}(W_n,x)\}\, W_n(x)\|_{L_p[-1,1]}$$

$$= \|\{P(x) - p_{mp}(w,x)\}w(x)\|_{L_p[-1,1]}$$

$$\le (1 - 1/A)\, \|Pw\|_{L_p[-1,1]} + \|\{P(x)/A - p_{mp}(w,x)\}w(x)\|_{L_p[-1,1]}$$

$$\le O(n^{-2}) + \psi(\|Pw\|_{L_p[-1,1]}/A) \qquad \text{(by (14.2) and (14.12))}$$

$$\le O(n^{-2}) + \psi(1/A) = O(n^{-\delta}), \tag{14.13}$$

some $\delta > 0$, by the definitions (14.3) and (14.4) of ψ.

<u>Step 2</u> From W_n to $V_{p,n}$.

We now replace $p_{mp}(W_n,x)$ in (14.13) by $p_{mp}(V_{p,n},x)$ for a suitable $V_{p,n}$. By Theorem 10.1 (with $\kappa := 1 - 1/p$), we can find polynomials S_n^- of degree at most $2(n - 2Cn\Delta_n - 2)$, positive in $[-1,1]$, except possibly for simple zeros at ± 1, such that

$$W_n^2(t) \le (1 - t^2)^{1-1/p}/S_n^-(t) =: V_{p,n}^2(t),\ t \in [-1,1]; \tag{14.14}$$

$$|W_n^2(t)/V_{p,n}^2(t) - 1| \le n^{-\delta},\ |t| \le 1 - n^{-\delta}; \tag{14.15}$$

and as $n \to \infty$, for some $\delta > 0$,

$$\int_{-1}^{1} |\log [W_n^2(t)/V_{p,n}^2(t)]|\ (1 - t^2)^{-1/2}\, dt = O(n^{-\delta}). \tag{14.16}$$

We now apply Lemma 14.1 with

$$P(x) := p_{mp}(V_{p,n},x) \text{ and } w := W_n.$$

Again, with the notation (14.1),

$$A = A(m) = E_{mp}(V_{p,n})^{-1}\, E_{mp}(W_n)$$

$$= E_{mp}(V_{p,n})^{-1}\, E_{mp}(W)\, c_n^{-m-1/p}\, (1 + O(n^{-2})) = 1 + O(n^{-\delta_1}),$$

as $n \to \infty$, for some $\delta_1 > 0$, by exactly the same reasoning as in the proof of Theorem 12.1, using (11.3) and (14.16). Further,

$$\|Pw\|_{L_p[-1,1]} = \|p_{mp}(V_{p,n},x)W_n(x)\|_{L_p[-1,1]}$$

$$\le \|p_{mp}(V_{p,n},x)V_{p,n}(x)\|_{L_p[-1,1]} = 1.$$

Exactly as in Step 1, we can use (14.2) to deduce that

$$\|\{p_{mp}(V_{p,n},x) - p_{mp}(W_n,x)\}\ W_n(x)\|_{L_p[-1,1]} = O(n^{-\delta}),$$

as $n \to \infty$, for some $\delta > 0$. It is clear from the above steps that the order term is uniform in m satisfying (14.11). Together with (14.13), this ensures that

$$\|\{c_n^{1/p}\ p_{mp}(W,c_nx) - p_{mp}(V_{p,n},x)\}W_n(x)\|_{L_p[-1,1]} = O(n^{-\delta}), \tag{14.17}$$

as $n \to \infty$, for some $\delta > 0$.

Step 3 From $p_{mp}(V_{p,n},x)$ to Szegö's Function

Now by (11.2), (11.3) and (11.6),

$$p_{mp}(V_{p,n},x) = \sigma_p^{-1} \operatorname{Re}[z^m D^{-2}(V_{p,n}(\cos t)|\sin t|^{1/p};\tfrac{1}{z})],$$

where $x = \cos \theta \in [-1,1]$, $\theta \in [0,\pi]$ and $z = e^{i\theta}$. This is the only stage of the proof where we use (14.11), namely in ensuring that the degree of S_n^- satisfies the hypotheses of Lemma 11.1. Then recalling the definition (2.12) of $\hat{F}_n$, and with $z = e^{i\theta}$ when $x = \cos \theta$,

$$\left\| \left\{ p_{mp}(V_{p,n},x) - \sigma_p^{-1} \operatorname{Re}[z^m D^{-2}(\hat{F}_n(\cos t);\tfrac{1}{z})] \right\} W(c_n x) \right\|_{L_p[-1,1]}$$

$$\leq \sigma_p^{-1} \left\| \left\{ D^{-2}(V_{p,n}(\cos t)\,|\sin t|^{1/p};\tfrac{1}{z}) - D^{-2}(\hat{F}_n(\cos t);\tfrac{1}{z}) \right\} W(c_n x) \right\|_{L_p[-1,1]}$$

$$= \sigma_p^{-1} \left\| \left\{ D^{-2}(V_{p,n}(\cos t)/W(c_n \cos t);\tfrac{1}{z}) - 1 \right\} (1 - x^2)^{-1/(2p)} \right\|_{L_p[-1,1]}$$

$$= \sigma_p^{-1} \left\{ \int_0^\pi | D^{-2}(f_n;e^{i\theta}) - 1|^p \, d\theta \right\}^{1/p},$$

where,

$$f_n(\theta) := V_{p,n}(\cos \theta)/W(c_n \cos \theta) \geq 1, \ \theta \in [-\pi,\pi],$$

by (14.14). Next, for any $0 < \epsilon < 1$, Lemma 14.2 and $f_n \geq 1$ show that

$$\int_0^\pi | D^{-2}(f_n;e^{i\theta}) - 1|^p \, d\theta$$

$$\leq C_1 \left\{ 2 \int_{\mathscr{S}_n} d\theta + \int_{-\pi}^\pi | f_n(\theta)^{-1} - 1|^p \, d\theta + \epsilon^p \, 2\pi \right\},$$

where

$$\operatorname{meas}(\mathscr{S}_n) \leq C_2 \, \epsilon^{-1} \int_{-\pi}^\pi | \log f_n(\theta)| \, d\theta.$$

Here C_1 and C_2 are independent of n and ϵ. Choosing $\epsilon := n^{-\delta/2}$, and using (14.15) to (14.16), we obtain

$$\int_0^\pi | D^{-2}(h_n;e^{i\theta}) - 1|^p \, d\theta \leq C_3 \, n^{-\delta_1}, \ n \to \infty,$$

for some $\delta_1 > 0$. Then we have

$$\left\| \left\{ p_{mp}(V_{p,n},x) - \sigma_p^{-1} \operatorname{Re}[z^m D^{-2}(\hat{F}_n(\cos t);\tfrac{1}{z})] \right\} W(c_n x) \right\|_{L_p[-1,1]}$$

$$= O(n^{-\delta}),\ n \to \infty,$$

some $\delta > 0$. Together with (14.17), this yields (14.10). □

We shall apply Theorem 14.3 to deduce asymptotics in the plane of L_p extremal polynomials for $1 < p < 2$, but first need:

Lemma 14.4

Let $1 \le p \le \infty$, and let $w \in L_p[-1,1]$ be a non–negative function satisfying

$$\int_{-1}^{1} \log w(x)\ (1 - x^2)^{-1/2}\ dx > -\infty. \tag{14.18}$$

Let $P(x)$ be a polynomial of degree at most m. Let

$$F(t) := w(t)\ (1 - t^2)^{1/(2p)},\ t \in [-1,1]. \tag{14.19}$$

Then for $u \in \mathbb{C}\backslash[-1,1]$,

$$|P(u)/\varphi(u)^m|$$
$$\le \pi^{-1/p} \|Pw\|_{L_p[-1,1]} |D^{-2}(F(\cos t);\tfrac{1}{\varphi(u)})| / (1 - |\varphi(u)|^{-1}). \tag{14.20}$$

Proof

This essentially appears in [30,p.160], and in the above form is Lemma 5.1 in [15]. Nevertheless, we present the details. Suppose first $p < \infty$. If as usual, $x = \cos\theta$, $z = e^{i\theta}$,

$$\|Pw\|^p_{L_p[-1,1]} = \frac{1}{2}\int_{-\pi}^{\pi} |P(\cos\theta)\ w(\cos\theta)|^p\ |\sin\theta|\ d\theta$$

$$= \frac{1}{2}\int_{-\pi}^{\pi} |P(\tfrac{1}{2}[z + z^{-1}])\ F(\cos\theta)|^p\ d\theta$$

$$= \frac{1}{2}\int_{-\pi}^{\pi} |S(z)\ D^2(F(\cos t);z)|^p\ d\theta,$$

where

$$S(z) := z^m P(\tfrac{1}{2}[z + z^{-1}])$$

is a polynomial of degree at most 2m in z. Next, by Cauchy's integral formula for $|s| < r < 1$,

$$|S(s)\, D^2(F(\cos t);s)| = \left|\frac{1}{2\pi i}\int_{|u|=r} \frac{S(u)\, D^2(F(\cos t);u)}{u - s}\, du\right|$$

$$\leq (r - |s|)^{-1} \left\{\frac{1}{2\pi}\int_{-\pi}^{\pi} |S(re^{i\theta})\, D^2(F(\cos t);re^{i\theta})|^p\, d\theta\right\}^{1/p} \left\{\frac{1}{2\pi}\int_{-\pi}^{\pi} d\theta\right\}^{1/q},$$

by Hölder's inequality, and where $1/q + 1/p = 1$. Letting $r \to 1-$, we obtain for $|s| < 1$,

$$|S(s)\, D^2(F(\cos t);s)| \leq (1 - |s|)^{-1}\, \pi^{-1/p}\, \|Pw\|_{L_p[-1,1]}.$$

Now let $u \in \mathbb{C}\backslash[-1,1]$ and set $s := 1/\varphi(u)$. Then

$$S(u) = \varphi(u)^{-m} P(\tfrac{1}{2}[\varphi(u)^{-1} + \varphi(u)]) = \varphi(u)^{-m} P(u),$$

and so (14.20) follows. The case $p = \infty$ follows by letting $p \to \infty$ in (14.20) and using continuity in p of the various terms. □

Theorem 14.5

Let $W \in SE(3)$. Let $1 < p < \infty$ and let $C_0 > 0$. Let $\{c_n\}_{n=1}^{\infty}$ be as in (2.11), with s fixed but large enough, and let $\{\hat{F}_n\}_{n=1}^{\infty}$ be as in (2.12). Then for some $\delta > 0$, as $n \to \infty$,

$$p_{mp}(W,c_n z)/\{c_n^{-1/p}\, \varphi(z)^m\, D^{-2}(\hat{F}_n(\cos t);\tfrac{1}{\varphi(z)})\}$$
$$= (2\sigma_p)^{-1}\, \{1 + O(n^{-\delta})\}, \tag{14.21}$$

uniformly for m in the range (14.11), and for z in any closed subset of $\mathbb{C}\backslash[-1,1]$.

Proof

For $m = n$, and $2 \leq p \leq \infty$, we already proved this in Theorem 12.2. Let $\{S_n^-\}_{n=1}^{\infty}$ and $\{V_{p,n}\}_{n=1}^{\infty}$ be exactly as in the proof of Theorem 14.3, satisfying (14.14) to (14.16). Recall too that (14.17) holds. By Lemma 14.4, for $u \in \mathbb{C}\backslash[-1,1]$,

$$|c_n^{1/2}\, p_{mp}(W,c_n u) - p_{mp}(V_{p,n},u)| / |\varphi(u)|^m$$

$$\leq \pi^{-1/p} \| \{c_n^{1/2}\, p_{mp}(W, c_n x) - p_{mp}(V_{p,n}, x)\}\, W(c_n x) \|_{L_p[-1,1]}$$

$$\times\, | D^{-2}(\hat{F}_n(\cos t); \tfrac{1}{\varphi(u)}) | \,/ (1 - 1/|\varphi(u)|)$$

$$\leq O(n^{-\delta})\, | D^{-2}(\hat{F}_n(\cos t); \tfrac{1}{\varphi(u)}) |, \tag{14.22}$$

uniformly for u in a closed subset of $\mathbb{C}\backslash[-1,1]$. Here δ is independent of the particular closed subset, and we have used (14.17). Next, Lemma 11.1 – in particular (11.8) – and our choice of the degree of S_n^- (in Theorem 14.3) show that

$$| p_{mp}(V_{p,n}, u) / \{ (2\sigma_p)^{-1} \varphi(u)^m D^{-2}(V_{p,n}(\cos t) | \sin t|^{1/p}; \tfrac{1}{\varphi(u)}) \} - 1 |$$

$$\leq | \varphi(u) |^{-C_3 n \Delta_n} \leq \exp(- C_4 n^{1/4}), \tag{14.23}$$

uniformly for u in closed subsets of $\mathbb{C}\backslash[-1,1]$. Furthermore, much as in the proof of Theorem 12.2, (see (12.13) to (12.14)), it follows from (14.16) that

$$D^{-2}(V_{p,n}(\cos t) | \sin t|^{1/p}; \tfrac{1}{\varphi(u)}) / D^{-2}(\hat{F}_n(\cos t); \tfrac{1}{\varphi(u)}) = 1 + O(n^{-\delta}),$$

uniformly for u in closed subsets of $\mathbb{C}\backslash[-1,1]$. Combining (14.22) to (14.23), we obtain that (14.21) holds uniformly for m satisfying (14.11) and for z in closed subsets of $\mathbb{C}\backslash[-1,1]$. □

<u>Proof of (2.10) in Theorem 2.2</u>

It remains to replace c_n by a_n in the special case m = n of (14.21). But by (12.15) and the remarks following it,

$$\frac{\varphi(a_n z/c_n)^n\, D^{-2}(\hat{F}_n(\cos t); 1/\varphi(a_n z/c_n))}{\varphi(z)^n\, D^{-2}(F_n(\cos t); 1/\varphi(z))} = 1 + O(n^{-\delta}), \tag{14.24}$$

as $n \to \infty$, uniformly for z in closed subsets of $\mathbb{C}\backslash[-1,1]$. Then replacing z by $a_n z/c_n$ in (14.21) (with m = n), and using (14.24), we obtain (2.10) also for $1 < p < 2$. □

15 Christoffel Functions

In this section, we establish Theorem 2.3, as a special case of Theorem 15.5 below. A more precise result is:

<u>Theorem 15.1</u>

Let $W \in SE$. Let $H := G_{Q/2}$ be defined by (1.15). Let $\{c_n\}_{n=1}^{\infty}$ be defined by (9.1) with s fixed but large enough. Let

$$\tilde{F}_n(x) := H(c_n x)^{-1} (1 - x^2)^{1/4},\ x \in (-1,1),\ n \geq 1. \tag{15.1}$$

Then there exist n_1 and $\delta > 0$ such that uniformly for $n \geq n_1$ and

$$|x| \leq 1 - n^{-\delta},\ x = \cos \phi, \tag{15.2}$$

we have

$$\begin{aligned}&\pi \lambda_n^{-1}(H^{-2}, c_n x)\, H^{-2}(c_n x)\, (1 - x^2)^{1/2}\, c_n/n \\ &= 1 - \frac{1}{2n} + \{2n \sin \phi\}^{-1} \sin((2n-1)\phi + 2\Gamma(\tilde{F}_n^2;\phi)) + n^{-1}\Gamma'(\tilde{F}_n^2;\phi) \\ &+ O(n^{-1-\delta}).\end{aligned} \tag{15.3}$$

<u>Proof</u>

We shall use Lemma 11.4, and the polynomials of Theorem 9.1. There will be three steps:

<u>Step 1</u> Lower Bounds

Let s in the definition of $\{c_n\}_{n=1}^{\infty}$ be chosen so large that

$$\|P/H\|_{L_2(\mathbb{R})} \leq (1 + n^{-2})\, \|P/H\|_{L_2(|x| \leq c_n)}, \tag{15.4}$$

$P \in \mathcal{P}_n$, $n \geq n_1$, this being possible by Lemma 13.5(a). For $n \geq n_1$, let $P_n(u)$ be the polynomial of Theorem 9.1 (with $\kappa = 1/2$) of degree at most 2n–4, positive in $[-1,1]$, except possibly for simple zeros at ± 1, and with the following properties: Let

$$\tau_n(u) := P_n(u)\, H^{-2}(c_n u)\, (1-u^2)^{-1/2},\ u \in (-1,1).$$

Then there exist n_1 and $\delta_1 > 0$ such that if $n \geq n_1$ and $|x_0| \leq 1 - n^{-\delta_1}$, then

$$\tau_n(x_0) = 1;\ \tau_n'(x_0) = 0, \tag{15.5}$$

while

$$\tau_n(u) \leq 1,\ u \in [-1,1]. \tag{15.6}$$

Furthermore (9.7) to (9.9) are valid. Then

$$H^{-2}(c_n u) \leq (1-u^2)^{1/2}/P_n(u) =: \Omega_n(u),\ u \in [-1,1], \tag{15.7}$$

and

$$H^{-2}(c_n x_0) = \Omega_n(x_0). \tag{15.8}$$

Next, using (15.4),

$$\lambda_n(H^{-2}, c_n x_0) \leq (1+n^{-2})^2 \inf_{P \in \mathcal{P}_{n-1}} \int_{-c_n}^{c_n} (P/H)^2(t)\, dt/P^2(c_n x_0)$$

$$\leq (1+n^{-2})^2\, c_n \inf_{R \in \mathcal{P}_{n-1}} \int_{-1}^{1} R^2(s) H^{-2}(c_n s)\, dt/R^2(x_0)$$

$$\leq (1+n^{-2})^2\, c_n \inf_{R \in \mathcal{P}_{n-1}} \int_{-1}^{1} R^2(s) \Omega_n(s)\, dt/R^2(x_0)$$

$$= (1+n^{-2})^2\, c_n\, \lambda_n(\Omega_n, x_0). \tag{15.9}$$

Next, if

$$g_n(x) := \Omega_n(x)\, (1-x^2)^{1/2} = (1-x^2)/P_n(x),\ x \in (-1,1),$$

Lemma 11.4 shows that for $x = \cos\phi \in (-1,1)$,

$$\pi\, \lambda_n^{-1}(\Omega_n, x)\, (1-x^2)^{1/2}\, \Omega_n(x) = n - \tfrac{1}{2}$$

$$+ \{2 \sin\phi\}^{-1} \sin((2n-1)\phi + 2\Gamma(g_n;\phi)) + \Gamma'(g_n;\phi). \tag{15.10}$$

From (15.8) and (15.9), we then obtain for $|x_0| = |\cos\phi_0| \leq 1 - n^{-\delta_1}$,

$$\pi\,\lambda_n^{-1}(H^{-2},c_n x_0)\,H^{-2}(c_n x_0)\,(1-x_0^2)^{1/2}\,c_n$$

$$\geq \pi\,\lambda_n^{-1}(\Omega_n,x_0)\,\Omega_n(x_0)\,(1-x_0^2)^{1/2}\,(1+n^{-2})^{-2}$$

$$= \{n - \tfrac{1}{2} + (2\sin\phi_0)^{-1}\sin((2n-1)\phi_0 + 2\Gamma(g_n;\phi_0))$$

$$+ \Gamma'(g_n;\phi_0)\}\,(1+n^{-2})^{-2}. \tag{15.11}$$

Next, from (15.1),

$$\tilde{F}_n^2(x)/g_n(x) = H^{-2}(c_n x)\,P_n(x)\,(1-x^2)^{-1/2} = \tau_n(x),$$

so that (compare (1.21)),

$$\Gamma(\tilde{F}_n^2;\phi) - \Gamma(g_n;\phi) = \Gamma(\tau_n;\phi). \tag{15.12}$$

To estimate $\Gamma'(\tau_n;\phi_0)$ and $\Gamma(\tau_n;\phi_0)$, we use Lemma 11.3. Firstly, from (11.23),

$$|\Gamma(\tau_n;\phi_0)| \leq \frac{(1-x_0^2)^{1/2}}{2\pi}\int_{-1}^{1}\left|\frac{\log\,\tau_n(s)-\log\,\tau_n(x_0)}{s-x_0}\right|(1-s^2)^{-1/2}\,ds$$

$$= O(n^{-\delta_1}\log n),$$

from (9.7) and (9.8), by a straightforward estimation and provided δ_1 is small enough. The estimation of $\Gamma'(\tau_n;\phi_0)$ is harder, and uses Lemma 11.3(d). Choose there $\epsilon := n^{-\delta_1/2}$, to obtain

$$|\Gamma'(\tau_n;\phi_0)| \leq n^{\delta_1/2}\pi^{-1}\int_{-1}^{1}\left|\frac{\log\,\tau_n(s)-\log\,\tau_n(x_0)}{s-x_0}\right|(1-s^2)^{-1/2}ds$$

$$+ 2|\,\tau_n'(x_0)/\tau_n(x_0)|\,\log(2en^{\delta_1/2}) + (1-|x_0|)^{1/2}\int_0^{n^{-\delta_1/2}}\omega(\eta)/\eta\,d\eta.$$

Here, according to Lemma 11.3(d), for $\eta > 0$,

$$\omega(\eta) = \sup\{\,|\tau_n'(s)/\tau_n(s) - \tau_n'(x_0)/\tau_n(x_0)| : |s-x_0| \leq \eta\}$$

$$= O(n^{-\delta_1}\eta),\ 0 \leq \eta \leq n^{-\delta_1},$$

by (9.7) with $j = 2$, and by (15.5). Hence, using (9.7) and (15.5), we obtain

$$|\Gamma'(\tau_n;\phi_0)| = O(n^{-\delta_1/2}\log n),$$

uniformly for $|x_0| = |\cos \phi_0| \le 1 - 2n^{-\delta_1/2}$, and n large enough. If δ_1 is small enough, (15.11) and (15.12) then yield for $|x_0| \le 1 - n^{-\delta_1}$,

$$\pi \lambda_n^{-1}(H^{-2}, c_n x_0) H^{-2}(c_n x_0) (1 - x_0^2)^{1/2} c_n$$
$$\ge (1 + n^{-2})^{-2} \{n - \tfrac{1}{2} + (2\sin \phi_0)^{-1}\sin((2n-1)\phi_0 + 2\Gamma(\tilde{F}_n^2;\phi_0))$$
$$+ \Gamma'(\tilde{F}_n^2;\phi_0) + O(n^{-\delta_1})\}. \tag{15.13}$$

We remark here that the δ_1 was suitably small enough to take care of the term $(2\sin(\phi_0))^{-1}$.

__Step 2__ $\Gamma'(\tilde{F}_n^2;\phi_0)$ is not so large.

To simplify (15.13), we show that $\Gamma'(\tilde{F}_n^2;\phi_0)$ is not too large, using Lemma 11.3(d), and if necessary, making δ_1 still smaller. Now from (15.1),

$$|\log \tilde{F}_n^2(x)| \le 2\,|\log H(c_n x)| + \tfrac{1}{2}\,|\log(1 - x^2)|$$
$$\le C_1\{Q(c_n x) + \log T(c_n x)\} + \tfrac{1}{2}\,|\log(1 - x^2)|$$
(by Theorem 6.1(a))
$$\le C_2\, n + \tfrac{1}{2}\,|\log(1 - x^2)|,$$

$|x| \le 1$, by Lemma 3.2(a) and (3.34) in Lemma 3.3(d). Also, by Theorem 6.1(a),

$$|\tfrac{d}{dx} \log \tilde{F}_n(x)| \le c_n\, |H'(c_n x)/H(c_n x)| + |x|/|1 - x^2|$$
$$\le C_3\{c_n Q'(c_n) + (1 - |x|)^{-1}\}$$
$$\le C_4\, n^{30/29},\ |x| \le 1 - n^{-\delta_2}, \text{ say,}$$

by Lemma 3.3(a) and (3.34) in Lemma 3.3(d). Similarly, using Theorem 6.1(a), we obtain for some C_5 and C_6,

$$|(\tfrac{d}{dx})^2 \log \tilde{F}_n(x)| \le C_5\, n^{C_6},\ |x| \le 1 - n^{-\delta_2}, \text{ say.}$$

Then choosing $\epsilon := n^{-1/29}$ in Lemma 11.3(d), we obtain

$$|\Gamma'(\tilde{F}_n^2;\phi_0)| \le C_7\, n^{30/29} \log n,\ |x_0| \le 1 - n^{-\delta_2}.$$

Hence, we can rewrite (15.13) as

$$\pi\,\lambda_n^{-1}(H^{-2},c_nx_0)\;H^{-2}(c_nx_0)\;(1-x_0^2)^{1/2}\;c_n$$

$$\geq n-\tfrac{1}{2}+(2\sin\,\phi_0)^{-1}\sin((2n-1)\phi_0+2\Gamma(\tilde{F}_n^2;\phi_0))$$

$$+\,\Gamma'(\tilde{F}_n^2;\phi_0)+O(n^{-\delta_1}), \tag{15.14}$$

for $|x_0| = |\cos\,\phi_0| \leq 1-n^{-\delta_2}$, and some small enough $\delta_2 > 0$.

Step 3 Upper Bounds

The proof of the upper bound corresponding to (15.14) is somewhat easier, since we do not need to use an infinite–finite range inequality.

For $n \geq n_1$, let $\hat{P}_n(u)$ be the polynomial of Theorem 9.1 (with $\kappa = 1/2$) of degree at most 2n–4, positive in [–1,1], except possibly for simple zeros at ± 1, and with the following properties: Let

$$\tau_n(u) := \hat{P}_n(u)\;H^{-2}(c_nu)\;(1-u^2)^{-1/2},\; u \in (-1,1).$$

Then (15.5) holds for a given $|x_0| \leq 1-n^{-\delta_1}$, (some fixed $\delta_1 > 0$), and

$$\tau_n(u) \geq 1,\; u \in [-1,1],$$

while (9.7) to (9.9) hold. Then

$$H^{-2}(c_nu) \geq (1-u^2)^{1/2}/\hat{P}_n(u) =: \hat{\Omega}_n(u),\; u \in (-1,1),$$

and

$$H^{-2}(c_nx_0) = \hat{\Omega}_n(x_0).$$

Then

$$\lambda_n(H^{-2},c_nx_0) \geq \inf_{P\in \mathcal{P}_{n-1}} \int_{-c_n}^{c_n} (P/H)^2(t)\;dt/P^2(c_nx_0)$$

$$= c_n \inf_{R\in \mathcal{P}_{n-1}} \int_{-1}^{1} R^2(s)H^{-2}(c_ns)\;dt/R^2(x_0)$$

$$\geq c_n \inf_{R\in \mathcal{P}_{n-1}} \int_{-1}^{1} R^2(s)\hat{\Omega}_n(s)\, dt/R^2(x_0)$$

$$= c_n\, \lambda_n(\hat{\Omega}_n,x_0).$$

Proceeding exactly as in Step 1, we obtain using Lemma 11.3 and Lemma 11.4,

$$\pi\, \lambda_n^{-1}(H^{-2},c_n x_0)\, H^{-2}(c_n x_0)\, (1-x_0^2)^{1/2}\, c_n$$

$$\leq \pi\, \lambda_n^{-1}(\hat{\Omega}_n,x_0)\, \hat{\Omega}_n(x_0)\, (1-x_0^2)^{1/2}$$

$$= n - \frac{1}{2} + (2\sin\phi_0)^{-1}\sin((2n-1)\phi_0 + 2\Gamma(\tilde{F}_n^2;\phi_0))$$

$$+ \Gamma'(\tilde{F}_n^2;\phi_0) + O(n^{-\delta_1}),$$

for $|x_0| = |\cos\phi_0| \leq 1 - n^{-\delta_2}$. Together with (15.14), this yields the result. □

We shall need

Lemma 15.2

Let $W \in SE(3)$ and let Q' be continuous at 0. Let $\{c_n\}_{n=1}^{\infty}$ be given by (9.1) with $s \geq 0$ fixed. Let $H := G_{Q/2}$, let Ψ be defined by (10.9), and for $n \geq 1$, let

$$Y_n := \max_{|u|\leq 1} \left|\frac{d}{du} \log \Psi(c_n u)\right|$$

$$= \max_{|u|\leq 1} |c_n H'(c_n u)/H(c_n u) - c_n Q'(c_n u)|. \tag{15.15}$$

Then

$$Y_n = o(n^{16/29}),\ n \to \infty. \tag{15.16}$$

If in addition, $W \in SE(4)$, then

$$Y_n = o(n^{3/29}),\ n \to \infty. \tag{15.17}$$

Proof

From Theorem 6.1(b), as $x \to \infty$,

$$H'(x)/H(x) - Q'(x) = O(Q'(x)/Q(x)^{1/2})$$
$$= O(\{Q'(x)\ T(x)/x\}^{1/2}),$$

by (6.2) or (2.4). Then by Lemma 3.1(b) and monotonicity of T, we obtain

$$Y_n = O(\{c_n Q'(c_n)\ T(c_n)\}^{1/2})$$
$$= o(n^{16/29}),$$

by Lemma 3.3(a) and by (3.34). If in addition $W \in SE(4)$, Theorem 6.1(c) ensures that as $x \to \infty$,

$$H'(x)/H(x) - Q'(x) = O(Q'(x)/Q(x)) = O(T(x)/x),$$

by (6.2). Then there exists $K > 0$ such that uniformly for $K/c_n \leq |u| \leq 1$,

$$c_n[H'(c_n u)/H(c_n u) - Q'(c_n u)] = O(c_n T(c_n u)/(c_n u))$$
$$= O(c_n T(c_n)) = o(n^{3/29}),$$

by (3.25) and (3.19). For $|u| \leq K/c_n$, (3.19) yields the result. □

The next step in obtaining asymptotics for the Christoffel functions $\lambda_n(W^2,x)$ is to replace c_n by a_n and $\tilde{F}_n$ by F_n in (15.3). Unfortunately, this leads to a loss of precision.

Theorem 15.3

Let $W \in SE(3)$, and let Q' be continuous at 0, satisfying near there a Lipschitz condition of some positive order. Let $H := G_{Q/2}$ be defined by (1.15), let μ_{n,a_n} be defined by (2.20), and let (as in (2.9) with $p = 2$),

$$F_n(x) := W(a_n x)\ (1 - x^2)^{1/4},\ x \in (-1,1),\ n \geq 1. \tag{15.18}$$

Then there exists $\delta > 0$ such that for

$$|x| \leq 1 - n^{-\delta},\ x = \cos\phi,\ z = e^{i\phi}, \tag{15.19}$$

we have

$$\pi\, \lambda_n^{-1}(H^{-2}, a_n x)\; H^{-2}(a_n x)\; (1 - x^2)^{1/2}\; a_n/n$$

$$= 1 + n^{-1}\Gamma'(F_n^2;\phi) + o(\epsilon_n), \tag{15.20}$$

$$= 1 + n^{-1}\; \mathrm{Re}[zD'(F_n^2(\cos t);z)/D(F_n^2(\cos t);z)] + o(\epsilon_n), \tag{15.21}$$

$$= \pi\, \mu_{n,a_n}(x)\; (1 - x^2)^{1/2} + o(\epsilon_n), \tag{15.22}$$

where

$$\epsilon_n := n^{-38/87},\; n \geq 1. \tag{15.23}$$

If in addition, $W \in SE(4)$, then we may choose

$$\epsilon_n := n^{-42/87},\; n \geq 1. \tag{15.24}$$

<u>Proof</u>

This will consist of two steps, each transforming (15.3).

<u>Step 1</u> From $\tilde{F}_n$ to $\hat{F}_n$.

We replace $\tilde{F}_n$ in (15.3) by $\hat{F}_n$, where

$$\hat{F}_n(x) := W(c_n x)\; (1 - x^2)^{1/4},\; x \in (-1,1), \tag{15.25}$$

and to this end, we use Lemma 11.3(d) to estimate

$$\Gamma'(\tilde{F}_n^2;\phi) - \Gamma'(\hat{F}_n^2;\phi) = \Gamma'(\tilde{F}_n^2/\hat{F}_n^2;\phi). \tag{15.26}$$

Note that from (15.1) and (15.25),

$$(\tilde{F}_n^2/\hat{F}_n^2)(x) = \{WG_{Q/2}\}^{-2}(c_n x) = \Psi^2(c_n x),$$

where Ψ is given by (10.9). Now for $|x| \leq 1$,

$$|\log \Psi(c_n x)| = O(\log T(c_n)) = O(\log n), \tag{15.27}$$

by Theorem 6.1(a) and Lemma 3.3(a). Also, for $|x| \leq 1$,

$$\frac{d}{dx} \log \Psi(c_n x) = O(Y_n), \tag{15.28}$$

by (15.15). Finally, it is not difficult to see from Theorem 6.1(a) that given $L > 0$, then for some C_3 and C_4,

$$\left|\left(\frac{d}{dx}\right)^2 \log \Psi(c_n x)\right| \leq C_3 n^{C_4},\quad L/c_n \leq |x| \leq 1. \tag{15.29}$$

Also as Q' satisfies a Lipschitz condition – say of order α – near 0, and since $G_{Q/2}$ is positive and entire, we see that for some C_5 and C_6,

$$\left|\frac{d}{dx}\log \Psi(c_n x) - \frac{d}{dy}\log \Psi(c_n y)\right| \le C_5 n^{C_6} |x-y|^{\alpha},\ |x|,\ |y| \le L/c_n. \tag{15.30}$$

Next, by Lemma 11.3(d), for given $\delta_2 > 0$,

$$|\Gamma'(\tilde{F}_n^2/\hat{F}_n^2;\phi)| \le n^{\delta_2}/\pi \int_{-1}^{1} \left|\frac{\log\ \Psi(c_n s)-\log\ \Psi(c_n x)}{s-x}\right| (1-s^2)^{-1/2}\, ds$$

$$+ 2\left|\frac{d}{dx}\log \Psi(c_n x)\right| \log(2en^{\delta_2}) + (1-|x|)^{1/2} \int_0^{n^{-\delta_2}} \omega(\eta)/\eta\, d\eta,$$

where for $\eta > 0$,

$$\omega(\eta) := \sup \left\{ \left|\frac{d}{dx}\log \Psi^2(c_n x) - \frac{d}{dt}\log \Psi^2(c_n t)\right| : |x-t| \le \eta\right\}.$$

Combining (15.27) to (15.30), we obtain

$$|\Gamma'(\tilde{F}_n^2/\hat{F}_n^2;\phi)| \le C_7 \{n^{\delta_2}(\log n)^2 + Y_n \log n\}$$

$$\le C_8 \log n \{n^{2\delta_2} + Y_n\},$$

provided

$$|x| = |\cos \phi| \le 1 - n^{-\delta_1},\ \text{some } \delta_1 > 0. \tag{15.31}$$

Note that δ_2 may be made as small as we please, while δ_1 depends on δ_2. Of course, we may assume that δ_1 is as small as we please (but fixed), since then the range (15.31) of x becomes smaller. Then from (15.3), we obtain uniformly for the range (15.31),

$$\pi \lambda_n^{-1}(H^{-2},c_n x)\, H^{-2}(c_n x)\, (1 - x^2)^{1/2}\, c_n/n$$

$$= 1 + n^{-1}\, \Gamma'(\hat{F}_n^2;\phi) + O\left(\frac{\log n}{n} \{n^{2\delta_2} + Y_n\}\right). \tag{15.32}$$

<u>Step 2</u> From c_n to a_n and $\hat{F}_n$ to F_n

We now make the substitution

$$c_n x = a_n u \tag{15.33}$$

in (15.32), using Lemma 11.5(b). Let us write

$$x = \cos \phi; \quad u = \cos \psi. \tag{15.34}$$

Note that in the notation (11.44), $\hat{F}_n$ and F_n become

$$\hat{F}_n = F_{c_n}; \quad F_n = F_{a_n}.$$

Thus we may apply Lemma 11.5(b) with $R_1 = a_n$; $R_2 = c_n$; $x_1 = u$; $x_2 = x$; $\phi_1 = \psi$; $\phi_2 = \phi$, and

$$\beta = a_n/c_n = (1 + s\Delta_n)^{-1}, \tag{15.35}$$

to deduce that

$$\begin{aligned}
&|\Gamma'(F_n^2;\psi) - \Gamma'(\hat{F}_n^2;\phi)| \\
&\leq \frac{4}{\pi}(1-\beta)\int_0^\beta \left|\frac{c_n sQ'(c_n s) - c_n xQ'(c_n x)}{s^2 - x^2}\right| (1-s^2)^{-1/2}\, ds \\
&+ \frac{2}{\pi}\int_\beta^1 \left|\frac{c_n sQ'(c_n s) - c_n xQ'(c_n x)}{s^2 - x^2}\right| (1-s^2)^{1/2}\, ds \\
&\leq C_9 \{\Delta_n \int_0^\beta \left|\frac{c_n sQ'(c_n s) - c_n xQ'(c_n x)}{c_n s - c_n|x|}\right| \frac{c_n}{s+|x|} (1-s^2)^{-1/2}\, ds \\
&+ \max_{s\in[\beta,1]} \left|\frac{c_n sQ'(c_n s) - c_n xQ'(c_n x)}{c_n s - c_n|x|}\right| c_n \Delta_n^{3/2}\}.
\end{aligned} \tag{15.36}$$

Now for some ξ between $|x|$ and s,

$$\left|\frac{c_n sQ'(c_n s) - c_n xQ'(c_n x)}{c_n s - c_n|x|}\right| = \left|\frac{d}{du}(uQ'(u))_{|u=c_n\xi}\right|$$

$$= T(c_n\xi)\, Q'(c_n\xi) \leq C_{10}\, T(c_n)Q'(c_n),$$

by monotonicity of T(t), and of Q'(t) for t large, and by continuity of Q' at 0. Then if also $x \neq 0$,

$$|\Gamma'(F_n^2;\psi) - \Gamma'(\hat{F}_n^2;\phi)| \leq C_{11}\{\Delta_n c_n Q'(c_n)T(c_n) \int_0^\beta \frac{1}{s+|x|} (1-s^2)^{-1/2}\, ds$$

$$+ T(c_n)Q'(c_n)c_n\ \Delta_n^{3/2}\}$$

$$\le C_{12}\ c_nQ'(c_n)\ T(a_{2n})^{1/3}\left(\frac{\log n}{n}\right)^{2/3}\ \{\log[2 + \frac{1}{|x|}] + \Delta_n^{1/2}\}$$

(by definition (1.12) of Δ_n and by (3.44))

$$\le C_{13}\ n^{34/87}\ (\log n)^{5/3}, \tag{15.37}$$

by (3.24), (3.25), provided $|x| \ge n^{-100}$. If $|x| \le n^{-100}$, then we need estimate only the integral in the right–hand side of (15.36) a little differently. As above, the integral over the range $[n^{-100},\beta]$ admits the estimate (15.37). We must estimate the integral over $[0,n^{-100}]$. First, recall that since Q is even and Q' is continuous at 0, so Q'(0) = 0. Also Q' satisfies a Lipschitz condition of order say α, so

$$|Q'(s)| \le C_{14}\ |s|^{\alpha},\ |s| \le C_{15}.$$

Then as $c_n n^{-\rho} = o(1)$ for each $\rho > 0$, and as tQ'(t) is even, positive, and increasing in $(0, \infty)$,

$$\Delta_n \int_{2|x|}^{2n^{-100}} \left|\frac{c_nsQ'(c_ns) - c_nxQ'(c_nx)}{c_ns - c_n|x|}\right| \frac{c_n}{s+|x|}\ (1-s^2)^{-1/2}\ ds$$

$$\le \Delta_n \int_{2|x|}^{2n^{-100}} \frac{c_nsQ'(c_ns)}{c_ns/2}\ \frac{c_n}{s+|x|}\ (1-s^2)^{-1/2}\ ds$$

$$\le \Delta_n \int_{2|x|}^{2n^{-100}} C_{16}\ (c_ns)^{\alpha-1}\ c_n^2\ ds = o(\Delta_n),$$

and, as above,

$$\Delta_n \int_0^{2|x|} \left|\frac{c_nsQ'(c_ns) - c_nxQ'(c_nx)}{c_ns - c_n|x|}\right| \frac{c_n}{s+|x|}\ (1-s^2)^{-1/2}\ ds$$

$$\le C_{17}\ \Delta_n\ T(c_n)Q'(c_n)\ c_n\ \frac{2|x|}{|x|}$$

$$\le C_{18}\ n^{34/87}\ (\log n)^{2/3},$$

as above. Hence (15.37) persists for $|x| \le n^{-100}$. Combining (15.32) (with c_nx

replaced by $a_n u$) and (15.37) leads to

$$\pi \lambda_n^{-1}(H^{-2}, a_n u)\, H^{-2}(a_n u)\, (1 - (a_n u/c_n)^2)^{1/2}\, c_n/n$$

$$= 1 + n^{-1}\, \Gamma'(F_n^2;\psi) + O(n^{-43/87}(\log n)^{5/3})$$

$$+ O(\frac{\log n}{n}\{n^{2\delta_2} + Y_n\}), \tag{15.38}$$

for the range (15.31) of $x = a_n u/c_n$. Now since $a_n/c_n = 1 + O(\Delta_n)$,

$$(1 - (a_n u/c_n)^2)^{1/2}/(1 - u^2)^{1/2} = 1 + O(\Delta_n(1 - u^2)^{-1})$$

$$= 1 + O(\Delta_n\, n^{\delta_1}),$$

for the range (15.31) of $x = a_n u/c_n$. Also, exactly as in step 2 of the proof of Theorem 15.1, we obtain from Lemma 11.3(d),

$$|\Gamma'(F_n^2;\psi)| \le C_{22}\, n^{30/29} \log n,$$

for the range (15.31) of $x = a_n u/c_n$, provided δ_1 is small enough. It follows that if δ_1, δ_2 are small enough, we may rewrite (15.38) as

$$\pi \lambda_n^{-1}(H^{-2}, a_n u)\, H^{-2}(a_n u)\, (1 - u^2)^{1/2}\, a_n/n$$

$$= 1 + n^{-1}\, \Gamma'(F_n^2;\psi) + O(n^{-43/87}(\log n)^{5/3}) + O(\frac{\log n}{n} Y_n), \tag{15.39}$$

uniformly for

$$|u| \le 1 - n^{-\delta_1},\ u = \cos\psi, \tag{15.40}$$

where δ_1 may be made arbitrarily small. Now by the estimates for Y_n in Lemma 15.2,

$$O(n^{-43/87}(\log n)^{5/3}) + O(\frac{\log n}{n} Y_n) = \begin{cases} o(n^{-38/87}), & W \in SE(3), \\ o(n^{-42/87}), & W \in SE(4). \end{cases}$$

Finally, by (11.45) in Lemma 11.5(a),

$$1 + n^{-1}\Gamma'(F_n^2;\psi)$$

$$= \pi (\sin\psi)\, \mu_{n,a_n}(\cos\psi) + \frac{1}{2n}$$

$$= 1 + n^{-1}\mathrm{Re}[e^{i\psi} D'(F_n^2(\cos t);e^{i\psi})/D(F_n^2(\cos t);e^{i\psi})],$$

by (11.18). On changing u to x, and ψ to ϕ, we have the stated result. □

To turn the asymptotics for $\lambda_n(H^{-2},a_n x)$ into those for $\lambda_n(W^2,x)$, we shall use Nevai's operators [21,6,8]. Recall that if, for example, $H = 1/G_{Q/2}$,

$$K_n(H^{-2},x,t) = \sum_{j=0}^{n-1} p_j(H^{-2},x)\, p_j(H^{-2},t)$$

$$= (\gamma_{n-1}/\gamma_n)(H^{-2})\, \frac{p_n(H^{-2},x)p_{n-1}(H^{-2},t) - p_n(H^{-2},t)p_{n-1}(H^{-2},x)}{x - t}. \tag{15.41}$$

We define for continuous $f:\mathbb{R} \to \mathbb{R}$ for which $f(x)|x|^j/H(x) \in L_1(\mathbb{R})$, $j = 0, 1, 2, \ldots$,

$$\mathcal{G}_n[f](x) := \lambda_n(H^{-2},x) \int_{-\infty}^{\infty} f(t)\, K_n^2(H^{-2},x,t)\, H^{-2}(t)\, dt, \tag{15.42}$$

$x \in \mathbb{R}$, $n \geq 1$. Nevai [21] introduced and studied these operators for weights on $[-1,1]$. For weights on $\mathbb{R}$, $\mathcal{G}_n$ and some generalisations were studied in [6] and applied to Christoffel functions in [8]. Here we follow [8]. A fundamental inequality of Nevai [21,p.76,Thm.6.2.3] asserts that if g, g^{-1} are positive and continuous, then (subject to finiteness),

$$1/\mathcal{G}_n[g^{-1}](x) \leq \lambda_n(gH^{-2},x)/\lambda_n(H^{-2},x) \leq \mathcal{G}_n[g](x),\ x \in \mathbb{R}. \tag{15.43}$$

We shall need some estimates:

<u>Lemma 15.4</u>

Let $W \in SE$, and let Q' be continuous at 0, satisfying there a Lipschitz condition of some positive order. Let $H = G_{Q/2}$ be defined by (1.15).

(a) There exists $\delta > 0$ such that

$$(1-x^2)^{1/4}\,(1-t^2)^{1/4}\,|K_n(H^{-2},a_n x,a_n t)|\; H^{-1}(a_n x)H^{-1}(a_n t)$$

$$\leq C_1/\{a_n|x - t|\},\quad |x|, |t| \leq 1 - n^{-\delta}. \tag{15.44}$$

Further, for $t \in \mathbb{R}$, and $|x| \leq 1 - n^{-\delta}$,

$$(1-x^2)^{1/4} \, |K_n(H^{-2},a_n x,a_n t)| \, H^{-1}(a_n x)$$

$$\leq C_2 \{ \sum_{j=n-1}^{n} |p_j(H^{-2},a_n t)| \}/\{ a_n^{1/2} \, |x-t| \}. \tag{15.45}$$

(b) Given $\epsilon > 0$, there exists $\delta > 0$ such that

$$C_3 \, a_n \, n^{-1+\epsilon} \geq \lambda_n(H^{-2},a_n x) \, H^2(a_n x) \geq C_4 \, a_n \, n^{-30/29}, \tag{15.46}$$

for $|x| \leq 1 - n^{-\delta}$.

(c) There exists $\delta > 0$ such that

$$|K_n(H^{-2},a_n x,a_n t)| \, H^{-1}(a_n x)H^{-1}(a_n t) \leq C_5 \, n^{30/29}/a_n, \tag{15.47}$$

$|x|, |t| \leq 1 - n^{-\delta}$.

<u>Proof</u>

(a) Recall first the elementary identity

$$(\gamma_{n-1}/\gamma_n)(H^{-2}) = \int_{-\infty}^{\infty} x p_{n-1}(H^{-2},x) \, p_n(H^{-2},x) \, H^{-2}(x) \, dx.$$

Applying Cauchy–Schwarz' inequality, and then Lemma 13.5(a) yields

$$(\gamma_{n-1}/\gamma_n)(H^{-2}) \leq \{\int_{-\infty}^{\infty} x^2 p_{n-1}^2(H^{-2},x)H^{-2}(x) \, dx\}^{1/2}$$

$$\leq (1 + n^{-2}) \, c_n \, \{\int_{-c_n}^{c_n} p_{n-1}^2(H^{-2},x)H^{-2}(x) \, dx\}^{1/2} \leq 2c_n. \tag{15.48}$$

Here c_n is defined by (9.1) with s large enough. (Of course, Theorem 9.1 implies a much more precise inequality, which we shall not need.) Next, according to Theorem 13.7,

$$|c_n^{1/2} \, p_j(H^{-2},c_n t) \, H^{-1}(c_n t) \, (1-t^2)^{1/4}| \leq 2, \tag{15.49}$$

for $j = n-1$, n, and $|t| \leq 1 - n^{-\delta_2}$, some $\delta_2 > 0$. Then (15.41), (15.48) and (15.49) imply (15.44) and (15.45).

(b) From Theorem 15.3,

$$\lambda_n^{-1}(H^{-2},a_n x)\, H^{-2}(a_n x)\, a_n/n$$
$$= \mu_{n,a_n}(x) + o(n^{-38/87}(1-|x|)^{-1/2}), \tag{15.50}$$

uniformly for $|x| \le 1 - n^{-\delta}$, some $\delta > 0$. Here, from Lemma 4.3(c), there exists $L > 0$ such that

$$\mu_{n,a_n}(x) \sim 1,\ L/a_n \le |x| \le 1/2,$$

and from Lemma 4.3(b), (a), for $1/2 \le |x| \le 1$,

$$\mu_{n,a_n}(x) = \nu_{n,a_n}(x)$$
$$\sim \frac{a_n}{n}\, \psi_{n,a_n}(x)\, (1 - |x|)^{1/2}$$
$$\ge \frac{a_n}{n}\, \psi_{n,a_n}(\tfrac{1}{2})\, (1 - |x|)^{1/2}$$
$$\sim \nu_{n,a_n}(1/2)\, (1 - |x|)^{1/2} \sim (1 - |x|)^{1/2}.$$

For $|x| \le L/a_n$, we may easily prove from (2.20) that $\mu_{n,a_n}(x) \sim 1$, using the Lipschitz condition of some positive order on Q'. Thus if $|x| \le 1 - n^{-\delta}$,

$$\mu_{n,a_n}(x) \ge C_1\, n^{-\delta/2}.$$

Also, by Lemma 4.3(e), for $1/2 \le |x| \le 1$,

$$\mu_{n,a_n}(x) \le C_2\, T(a_n)^{1/2} \le C_3\, n^{1/29}.$$

Then these estimates and (15.50) yield (15.46).

(c) By the Cauchy–Schwarz inequality,

$$K_n(H^{-2},a_n x,a_n t) \le \{\sum_{j=0}^{n-1} p_j^2(H^{-2},a_n x)\}^{\frac{1}{2}}\, \{\sum_{j=0}^{n-1} p_j^2(H^{-2},a_n t)\}^{\frac{1}{2}}$$
$$= \lambda_n^{-1/2}(H^{-2},a_n x)\, \lambda_n^{-1/2}(H^{-2},a_n t).$$

Then the lower bound in (15.46) yields (15.47). □

Finally we can prove Theorem 2.3:

Theorem 15.5

Let $W \in SE(3)$ and let Q' be continuous at 0, satisfying there a Lipschitz condition of some positive order. Let $\{F_n\}_{n=1}^{\infty}$ be defined by (15.18), and let $\{\mu_{n,a_n}\}_{n=1}^{\infty}$ be defined by (2.20). Then there exists $\delta > 0$ such that as $n \to \infty$, uniformly for

$$|x| \leq 1 - n^{-\delta};\ x = \cos\phi;\ z = e^{i\phi}, \tag{15.51}$$

we have

$$\pi\, \lambda_n^{-1}(W^2,a_nx)\ W^2(a_nx)\ (1-x^2)^{1/2}\ a_n/n$$

$$= 1 + n^{-1}\ \mathrm{Re}[zD'(F_n^2(\cos t);z)/D(F_n^2(\cos t);z)] + o(\epsilon_n), \tag{15.52}$$

$$= 1 + n^{-1}\Gamma'(F_n^2;\phi) + o(\epsilon_n), \tag{15.53}$$

$$= \pi\, \mu_{n,a_n}(x)\ (1-x^2)^{1/2} + o(\epsilon_n), \tag{15.54}$$

where

$$\epsilon_n := n^{-19/58},\ n \geq 1. \tag{15.55}$$

If in addition $W \in SE(4)$, then we may take

$$\epsilon_n := n^{-42/87},\ n \geq 1. \tag{15.56}$$

Proof

We let $H := G_{Q/2}$ for notational simplicity. Recall from (10.9), the definition of Ψ. Then Nevai's inequality (15.43) implies that

$$\{\mathcal{G}_n[\Psi^2](a_nx)/\Psi^2(a_nx)\}^{-1} \leq \frac{\lambda_n(W^2,a_nx)\ W^{-2}(a_nx)}{\lambda_n(G_{Q/2}^{-2},a_nx)\ G_{Q/2}^2(a_nx)}$$

$$\leq \mathcal{G}_n[\Psi^{-2}](a_nx)/\Psi^{-2}(a_nx). \tag{15.57}$$

Let $\sigma := \pm 2$, and let $\delta > 0$. Further, let $\{\theta_n\}_{n=1}^{\infty}$ be a strictly decreasing sequence of positive numbers with limit 0. Since $\mathcal{G}_n[1] = 1$, (see [8,21]), we have for

$$|x| \leq 1 - \theta_n - 2n^{-\delta}, \tag{15.58}$$

$$\mathcal{G}_n[\Psi^{\sigma}](a_nx)/\Psi^{\sigma}(a_nx) - 1$$

$$= \lambda_n(H^{-2},a_n x) \int_{-\infty}^{\infty} K_n^2(H^{-2},a_n x,a_n t) \left[\frac{\Psi^\sigma(a_n t)-\Psi^\sigma(a_n x)}{\Psi^\sigma(a_n x)}\right] a_n H^{-2}(a_n t)\, dt$$

$$= \lambda_n(H^{-2},a_n x) \left\{\int_{|x-t|\le\theta_n} + \int_{\substack{|x-t|\ge\theta_n \\ |t|\le 1-n^{-\delta}}} + \int_{|t|\ge 1-n^{-\delta}}\right\}$$

$$\times\, K_n^2(H^{-2},a_n x,a_n t) \left[\frac{\Psi^\sigma(a_n t)-\Psi^\sigma(a_n x)}{\Psi^\sigma(a_n x)}\right] a_n H^{-2}(a_n t)\, dt$$

$$=: I_1 + I_2 + I_3. \tag{15.59}$$

Now by (15.46) and (15.47),

$$I_1 \le C_6\, n^{31/29+\epsilon} \int_{|x-t|\le\theta_n} \left|\frac{\Psi^\sigma(a_n t)-\Psi^\sigma(a_n x)}{\Psi^\sigma(a_n x)}\right| dt. \tag{15.60}$$

Here, by choosing δ small enough, we may ensure that $\epsilon > 0$ is as small as we please. Now for x as in (15.58), and t such that $|x-t| \le \theta_n$,

$$\Psi^\sigma(a_n t)/\Psi^\sigma(a_n x) = \exp(\sigma \int_x^t \frac{d}{du} \log \Psi(a_n u)\, du)$$

$$= \exp\,(O(|x-t|\, Y_n)) = \exp(O(\theta_n Y_n)),$$

by (15.15). Thus if

$$\lim_{n\to\infty} \theta_n Y_n = 0, \tag{15.61}$$

then uniformly for x as in (15.58) and $|x-t| \le \theta_n$,

$$\Psi^\sigma(a_n t)/\Psi^\sigma(a_n x) = 1 + O(\theta_n Y_n).$$

Then (15.60) yields

$$I_1 \le C_7\, n^{31/29+\epsilon}\, \theta_n^2 Y_n. \tag{15.62}$$

Next, using (15.44) and (15.46),

$$I_2 \le C_8\, n^{-1+\epsilon}\, (1-x^2)^{-1/2}$$

$$\times \int_{\substack{|x-t|\geq \theta_n \\ |t|\leq 1-n^{-\delta}}} \frac{1}{|x-t|^2} \left| \frac{\Psi^\sigma(a_n t)-\Psi^\sigma(a_n x)}{\Psi^\sigma(a_n x)} \right| \frac{dt}{(1-t^2)^{1/2}}.$$

Here for some ξ between t and x, and then using (15.15),

$$\frac{\Psi^\sigma(a_n t)-\Psi^\sigma(a_n x)}{\Psi^\sigma(a_n x)} = \frac{(a_n t - a_n x)}{\Psi^\sigma(a_n x)} \frac{d}{du}\Psi^\sigma(u)|_{u=a_n\xi}$$

$$= O(|t-x| \frac{\Psi^\sigma(a_n\xi)}{\Psi^\sigma(a_n x)} Y_n) = O(|t-x| T(a_n)Y_n),$$

since $\sigma = \pm 2$, and since $\Psi \sim T^{-1/2}$ by Theorem 6.1(a). Thus

$$I_2 \leq C_8\, n^{-1+\epsilon+\delta/2}\, T(a_n)Y_n \log(1/\theta_n), \tag{15.63}$$

for x satisfying (15.58). Finally, using (15.45) and (15.46), we obtain for the range (15.58), since $|t| \geq 1-n^{-\delta}$ implies $|x-t| \geq n^{-\delta}$,

$$I_3 \leq C_9\, a_n n^{-1+\epsilon+5\delta/2} \sum_{j=n-1}^{n} \int_{|t|\geq 1-n^{-\delta}} p_j^2(H^{-2},a_n t)H^{-2}(a_n t)$$

$$\times \{\Psi^\sigma(a_n t)/\Psi^\sigma(a_n x) + 1\}\, dt.$$

Now by Lemma 13.5(b), for j = n–1, n, and n large enough,

$$\int_{\mathbb{R}} p_j^2(H^{-2},s)\, \Psi^\sigma(s)\, H^{-2}(s)\, ds$$

$$\leq 2 \int_{-c_{2n}}^{c_{2n}} p_j^2(H^{-2},s)\, \Psi^\sigma(s)\, H^{-2}(s)\, ds \leq 2\, \|\Psi^\sigma\|_{L_\infty(|x|\leq c_{2n})}.$$

Hence, for n large enough,

$$I_3 \leq C_{10}\, n^{-1+\epsilon+5\delta/2}\, T(a_{2n}) \leq C_{11}\, n^{-1+\epsilon+5\delta/2+2/29}, \tag{15.64}$$

by (3.25). Hence for the range (15.58), we obtain from (15.59), (15.62), (15.63) and (15.64),

$$|\mathcal{G}_n[\Psi^\sigma](a_n x)/\Psi^\sigma(a_n x) - 1|$$

$$\leq C_{12}\{n^{31/29+\epsilon}\, \theta_n^2\, Y_n + n^{-27/29+\epsilon+\delta} Y_n \log(1/\theta_n) + n^{-27/29+\epsilon+5\delta/2}\}.$$

Note from Lemma 15.2 that we may assume that Y_n grows at least as fast as some positive power of n (else replace it by the relevant upper bound of Lemma 15.2). Then choosing

$$\theta_n := \min\{ n^{-100}, Y_n^{-1} (\log n)^{-1}\},$$

we can satisfy (15.61), and for the range (15.58), then obtain

$$|\mathcal{G}_n[\Psi^\sigma](a_n x)/\Psi^\sigma(a_n x) - 1|$$
$$\leq C_{13}\{n^{-27/29+\epsilon+\delta} Y_n \log n + n^{-27/29+\epsilon+5\delta/2}\}.$$

Of course here ϵ and δ could be made as small as we please. Then from (15.16), if ϵ, δ are small enough,

$$|\mathcal{G}_n[\Psi^\sigma](a_n x)/\Psi^\sigma(a_n x) - 1|$$
$$\leq C_{14} n^{-11/29+\epsilon+\delta} \log n \leq C_{15}\, n^{-21/58}.$$

If in addition $W \in SE(4)$, then from (15.17), if ϵ, δ are small enough, we have for the range (15.58),

$$|\mathcal{G}_n[\Psi^\sigma](a_n x)/\Psi^\sigma(a_n x) - 1|$$
$$\leq C_{16}\, n^{-24/29+\epsilon+\delta} \log n \leq C_{17}\, n^{-47/58}. \qquad (15.65)$$

Finally, from (15.57) and from Theorem 15.3, for the range (15.58) of x,

$$\pi\, \lambda_n^{-1}(W^2, a_n x)\, W^2(a_n x)\, (1-x^2)^{1/2}\, a_n/n$$
$$= \{\pi\, \mu_{n,a_n}(x)\, (1-x^2)^{1/2} + o(n^{-38/87})\}\{1 + O(n^{-21/58})\}$$
$$= \pi\, \mu_{n,a_n}(x)\, (1-x^2)^{1/2} + O(T(a_{2n})^{1/2}\, n^{-21/58}) \quad \text{(by (4.38))}$$
$$= \pi\, \mu_{n,a_n}(x)\, (1-x^2)^{1/2} + o(n^{-19/58}),$$

by (3.25). If $W \in SE(4)$, then from (15.24) and (15.65), we replace $n^{-38/87}$ by $n^{-42/87}$ and $n^{-21/58}$ by $n^{-47/58}$ in the above, which leads to (15.56). □

References

1. **W.C.Bauldry**, Orthogonal Polynomials Associated with Exponential Weights, Ph.D. Dissertation, Ohio State University, Columbus, 1985.

2. **D.Braess**, Nonlinear Approximation Theory, Springer Series in Computational Mathematics, Vol. 7, Springer, Berlin, 1986.

3. **P.Erdős**, On the Distribution of Roots of Orthogonal Polynomials, (in) Proceedings of the Conf. Constr. Theory Fns., (G.Alexits, et al., eds.), pp.145–150, Akademiai Kiado, Budapest, 1972.

4. **G.Freud**, Orthogonal Polynomials, Akademiai Kiado/Pergamon Press, Budapest, 1971.

5. **J.B.Garnett**, Bounded Analytic Functions, Academic Press, New York, 1981.

6. **A.Knopfmacher**, Positive Convergent Approximation Operators Associated with Orthogonal Polynomials for Weights on the Whole Real Line, J. Approx. Theory, 46(1986), 182–203.

7. **A.Knopfmacher and D.S.Lubinsky**, Analogues of Freud's Conjecture for Erdös Type Weights and Related Polynomial Approximation Problems, (in) Approximation Theory, Tampa, (E.B. Saff,ed.),pp. 21–69, Springer, Berlin, 1987.

8. **A.Knopfmacher and D.S.Lubinsky**, Asymptotic Behaviour of the Ratio of Christoffel Functions for Weights W^2 and W^2g, J. Approx. Theory, 52(1988), 293–314.

9. **A.L.Levin and D.S.Lubinsky**, L_∞ Markov and Bernstein Inequalities for Freud Weights, submitted.

10. **G.Lopez and E.A.Rahmanov**, Rational Approximations, Orthogonal Polynomials and Equilibrium Distributions, to appear in Proc. Segovia 1986.

11. **D.S.Lubinsky**, Gaussian Quadrature, Weights on the Whole Real Line, and Even Entire Functions with Nonnegative Even Order Derivatives, J. Approx. Theory, 46(1986), 297–313.

12. **D.S.Lubinsky**, A Survey of General Orthogonal Polynomials for Weights on Finite and Infinite Intervals, Acta Applicandae Mathematicae, 10(1987), 237–296.

13. **D.S.Lubinsky**, L_∞ Markov and Bernstein Inequalities for Erdös Weights, to appear in J. Approx. Theory.

14. **D.S.Lubinsky and E.B.Saff**, Strong Asymptotics for Extremal Errors and Extremal Polynomials Associated with Weights on $(-\infty,\infty)$, Springer Lecture Notes in Math., Vol. 1305, Springer, Berlin, 1988.

15. **D.S.Lubinsky and E.B.Saff**, The Asymptotic Behaviour of L_p Extremal Errors and Polynomials $(1 \le p \le \infty)$ Associated with Weights on [–1,1], submitted.

16. **Al.Magnus**, On Freud's Equations for Exponential Weights, J. Approx. Theory, 46(1986), 65–99.

17. **A.Mâté,P.Nevai and V.Totik**, Extensions of Szegö's Theory of Orthogonal Polynomials, III, Constructive Approximation, 3(1987), 73–96.

18. **A.Mâté,P.Nevai and V.Totik**, Asymptotics for the Zeros of Orthogonal Polynomials Associated with Infinite Intervals, J. London Math. Soc., 33(1986), 303–310.

19. **H.N.Mhaskar and E.B.Saff**, Extremal Problems for Polynomials with Exponential Weights, Trans. Amer. Math. Soc., 285(1984), 203–234.

20. **H.N.Mhaskar and E.B.Saff**, Where Does The Sup Norm of a Weighted Polynomial Live ?, Constructive Approximation, 1(1985), 71–91.

21. **P.Nevai**, Orthogonal Polynomials, Memoirs Amer. Math. Soc., 213(1979).

22. **P.Nevai**, Asymptotics for Orthogonal Polynomials Associated with $\exp(-x^4)$, SIAM J. Math. Anal., 15(1984), 1177–1187.

23. **P.Nevai**, Geza Freud, Orthogonal Polynomials and Christoffel Functions, A Case Study, J.Approx. Theory, 48(1986), 3–167.

24. **P.Nevai and J.S.Dehesa**, On Asymptotic Average Properties of Zeros of Orthogonal Polynomials, SIAM J. Math. Anal., 10(1979), 1184–1192.

25. **P.Nevai and V.Totik**, Weighted Polynomial Inequalities, Constructive Approximation, 2(1986), 113–127.

26. **F.W.J.Olver**, Asymptotics and Special Functions, Academic Press, New York, 1974.

27. **E.A.Rahmanov**, On Asymptotic Properties of Polynomials Orthogonal on the Real Axis, Math USSR Sbornik, 47(1984), 155–193.

28. **E.A.Rahmanov**, On Asymptotic Properties of Polynomials Orthogonal on the Circle with Weights not Satisfying Szegö's Condition, Math. USSR Sbornik, 58(1987), 149–167.

29. **R.C.Sheen**, Plancherel–Rotach–Type Asymptotics for Orthogonal Polynomials Associated with $\exp(-x^6/6)$, J. Approx. Theory, 50(1987), 232–293.

30. **G.A.Szegö**, Orthogonal Polynomials, Amer. Math. Soc. Colloq. Publns., Vol. 23,

Amer. Math. Soc., Providence, R.I., 1939, 4th ed., 1975.

31. **W.Van Assche**, Asymptotics for Orthogonal Polynomials, Springer Lecture Notes in Mathematics, Vol. 1265, Springer, Berlin, 1987.

Appendix: Freud Weights

This research note built on earlier joint work with E. B. Saff [14] on Freud weights, that is weights of the form $W := e^{-Q}$, where Q is even and of polynomial growth at infinity. In this appendix, we use the methods of Sections 9 to 15 above, to obtain asymptotics for Christoffel functions and orthogonal polynomials on the real line.

Probably the main results of this appendix are the asymptotics for Christoffel functions for Freud weights, presented in Theorems I.2 and I.3. They largely solve an extended version of a conjecture of Nevai [23,p.143]. While the asymptotics for orthogonal polynomials are interesting and useful, it remains to be seen how they relate to the forthcoming results of E. A. Rahmanov, who in [10] announced asymptotics for the weights $\exp(-|x|^{\alpha})$, $\alpha > 0$.

The appendix is organised as follows: In Section I, we state our main results. In Section II, we present some elementary estimates. In Section III and IV, weighted and one–sided weighted approximations are constructed. In Section V, we establish asymptotics for orthogonal polynomials, and in Section VI, asymptotics for Christoffel functions.

We retain the notation of the main part of this monograph, and frequently refer to the lemmas above, especially when proofs are similar. The reader will need a copy of [14].

1. Main Results

First we define a suitable class of weights:

<u>Definition I.1</u>

(a) Let $W := e^{-Q}$, where $Q: \mathbb{R} \to \mathbb{R}$ is even; $Q(0) = 0$; Q'' is continuous and Q' is positive in $(0, \infty)$; and for some Λ, $\Omega > 0$,

$$\Lambda \leq (xQ'(x))'/Q'(x) \leq \Omega, \; x \in (0, \infty). \tag{I.1}$$

Then we shall call W a **Freud weight of type Λ**, and write $W \in FR_\Lambda$.

(b) Let $j \geq 3$. We write $W \in FR_\Lambda(j)$ if $W \in FR_\Lambda$ and $Q^{(j)}(x)$ exists for x large enough, while for $k = 3, 4, \ldots j$,

$$x^{j-1} \, | Q^{(j)}(x)| \, /Q'(x) \leq C_1, \; x \geq C_2. \tag{I.2}$$

Note that $\exp(-|x|^\alpha) \in FR_\Lambda(j)$ for all positive integers j, if $\alpha \geq \Lambda$, but $\exp(-|x|^\alpha) \notin FR_\Lambda$ if $\alpha < \Lambda$. Note too that the class $FR_\Lambda(3)$ coincides with the class of weights considered in Theorem 3.1 in [14,p.17].

Following is the main result of this appendix:

<u>Theorem I.2</u>

Let $W \in FR_\Lambda(3)$, some $\Lambda > 1$. Let $\eta > 0$, and let μ_{n,a_n} be defined by (2.20). Let

$$F_n(x) := W(a_n x) \, (1 - x^2)^{1/4}, \; x \in (-1,1), \; n \geq 1. \tag{I.3}$$

Then there exists $\delta > 0$, such that uniformly for

$$|x| \leq 1 - n^{-\delta}; \quad x = \cos \phi; \quad z = e^{i\phi}, \tag{I.4}$$

we have

$$\pi \lambda_n^{-1}(W^2, a_n x)\, W^2(a_n x)\, (1 - x^2)^{1/2}\, a_n/n$$

$$= 1 + n^{-1} \operatorname{Re}[zD'(F_n^2(\cos t);z)/D(F_n^2(\cos t);z)] + O(\epsilon_n), \tag{I.5}$$

$$= 1 + n^{-1} \Gamma'(F_n^2;\phi) + O(\epsilon_n), \tag{I.6}$$

$$= \pi \mu_{n,a_n}(x)\, (1 - x^2)^{1/2} + O(\epsilon_n), \tag{I.7}$$

where

$$\epsilon_n := n^{\eta - \min\{1/2,\, 1 - 1/\Lambda\}}, \quad n \geq 1. \tag{I.8}$$

If in addition $W \in FR_\Lambda(4)$, then we may choose

$$\epsilon_n := n^{\eta - \min\{1/2,\, 2/3 - 1/\Lambda\}}, \quad n \geq 1. \tag{I.9}$$

Proof

See Theorem VI.5. □

Remarks

(a) Usually Plancherel–Rotach asymptotics of the above type are given for the more restrictive range $|x| \leq 1 - \epsilon$, $\epsilon \in (0,1)$ fixed. If we consider only this more restrictive range, then the order terms in (I.8) to (I.9) can be slightly improved.

(b) The proof of Lemma VI.4 shows that

$$\mu_{n,a_n}(x) \sim (1 - |x|)^{1/2}, \tag{I.10}$$

uniformly for $|x| \leq 1$ and n large enough. Hence (I.7) yields applicable upper and lower bounds for $\lambda_n(W^2, a_n x)$ for n large enough.

(c) The above result establishes one form of a conjecture of Nevai [23, p.143] for more general weights, but with a weaker order estimate. Nevai considered $\exp(-|x|^\alpha)$, $\alpha > 1$, in which case $\mu_{n,a_n}(x)$ is the Nevai–Ullman measure, independent of n [19,p.207]:

$$\mu_{n,a_n}(x) = \alpha\pi^{-1} \int_{|x|}^{1} t^{\alpha-1} (t^2-x^2)^{-1/2}\, dt, \; x \in [-1,1].$$

Further, Nevai conjectured $\epsilon_n = O(n^{-1})$, whereas for $\exp(-|x|^{\alpha})$, $\alpha \geq 3$, (I.9) yields $\epsilon_n = O(n^{-2/3+\eta})$, $\eta > 0$ arbitrary. For weights such as $\exp(-x^4)$ or $\exp(-x^6)$, results of this type appeared in [1,22,23,29].

(d) Theorem I.2 (= Theorem VI.5) is deduced from a much more precise result, Theorem VI.1, which contains a second order term, without the restriction $\Lambda > 1$, but for weights that are the reciprocal of an entire function. Unfortunately a lot of accuracy is lost in passing from c_n (see (I.12) below) to a_n and from $1/G_{Q/2}$ to W. We state a special case of Theorem VI.1:

Theorem I.3

Let $W \in FR_{\Lambda}$, some $\Lambda > 0$, and assume that $1/W$ is an even entire function with non–negative Maclaurin series coefficients. Let $s > 0$ be fixed, but large enough, let

$$\Delta_n := (\frac{\log n}{n})^{2/3}, \; n \geq 2, \tag{I.11}$$

and let

$$c_n := a_n(1 + s\Delta_n), \; n \geq 2. \tag{I.12}$$

Let

$$\hat{F}_n(x) := W(c_n x)\, (1 - x^2)^{1/4}, \; x \in (-1,1), \; n \geq 2. \tag{I.13}$$

Then there exists $\delta > 0$ such that uniformly for the range (I.4),

$$\pi\, \lambda_n^{-1}(W^2, c_n x)\, W^2(c_n x)\, (1 - x^2)^{1/2}\, c_n/n$$

$$= 1 + n^{-1}\Gamma'(\hat{F}_n^2;\phi) - \frac{1}{2n} + \{2n \sin\phi\}^{-1} \sin((2n-1)\phi + 2\Gamma(\hat{F}_n^2;\phi))$$

$$+ O(n^{-1-\delta}), \tag{I.14}$$

$$= \pi\, \mu_{n,c_n}(x)\, (1 - x^2)^{1/2}$$

$$+ \{2n \sin\phi\}^{-1} \sin(2n\pi \int_x^1 \mu_{n,c_n}(t)\, dt - \pi/2) + O(n^{-1-\delta}), \tag{I.15}$$

where μ_{n,c_n} is defined by (4.3) to (4.5).

Proof

See Theorem VI.1 and set $H = 1/W$ there. Then $HW = 1$ and $(HW)^{(j)} \equiv 0$, $j = 1, 2$, and the conditions of Theorem III.1 are easily verified, using (I.1) and (II.2) below. The second form (I.15) for the asymptotic follows from (11.46). □

Remarks

(a) Since $c_n > a_n$, $\mu_{n,c_n}(x)$ is actually negative in a small neighbourhood of $x = 1$, but is of course positive for the range (I.4).

(b) The above result improves on Nevai's Conjecture in [23,p.143], providing a second order term for weights such as

$$W(x) := \exp(-x^{2k}), \text{ k a positive integer.}$$

However it applies to more general weights, including "reciprocal entire" weights that behave like $\exp(-|x|^{\alpha})$, $0 < \alpha < 1$, for which the corresponding moment problem is indeterminate. For such weights,

$$\lim_{n\to\infty} \lambda_n^{-1}(W^2,x) = \sum_{j=0}^{\infty} p_j^2(W^2,x) < \infty, \ x \in \mathbb{R},$$

and

$$\lim_{n\to\infty} c_n/n = \lim_{n\to\infty} a_n/n = \infty,$$

but this is consistent with (I.15), since $\mu_{n,c_n}(x)$ becomes unbounded near 0 as $n \to \infty$.

Following is our asymptotic for the orthogonal polynomials on the real axis:

Theorem I.4

Let $W \in FR_\Lambda(3)$, some $\Lambda > 3$. Let $s > 0$ be fixed, but large enough, and let $\{c_n\}_{n=1}^\infty$ and $\{\hat{F}_n\}_{n=1}^\infty$ be as in Theorem I.3. Then there exists $\delta > 0$, such that uniformly for the range (I.4), we have

$$c_n^{1/2}\, p_n(W^2,c_nx)\, W(c_nx)\, (1-x^2)^{1/4}$$

$$= (2/\pi)^{1/2}\, \mathrm{Re}[z^n\, \hat{F}_n(x)/\, D(\hat{F}_n^2(\cos\, t);z)] + O(n^{-\delta}), \tag{I.16}$$

$$= (2/\pi)^{1/2} \cos(n\phi + \Gamma(\hat{F}_n^2;\phi)) + O(n^{-\delta}), \tag{I.17}$$

$$= (2/\pi)^{1/2} \cos(n\pi \int_x^1 \mu_{n,c_n}(t)\, dt + \phi/2 - \pi/4) + O(n^{-\delta}). \tag{I.18}$$

The result remains valid if $W \in FR_\Lambda$, some $\Lambda > 0$, under the additional restriction that W^{-1} is an even entire function with non–negative Maclaurin series coefficients.

Proof

For $W \in FR_\Lambda(3)$, $\Lambda > 3$, this is the special case $m = n$ of Theorem V.5. When W^{-1} is entire, this is the special case $m = n$ of Theorem V.7 with $H := 1/W$ there. □

Remarks

(a) The restriction $\Lambda > 3$ in the first part of Theorem I.4 is unfortunate, and means, for example, that the result applies to $\exp(-|x|^\alpha)$ only if $\alpha > 3$ (the second part treats the classical Hermite case $\alpha = 2$ also). The restriction on Λ arises in passing from $1/G_{Q/2}$ to W, via the techniques of the relativized Szegö theory, and in forming suitable polynomial approximations. For $\Lambda \le 3$, we can apparently prove (I.16) to (I.18) for the range

$$n^{-1/3+\epsilon} \le |x| \le 1 - n^{-\delta},$$

instead of (I.4). Here $\epsilon > 0$ may be made arbitrarily small, while δ is some fixed small enough positive number.

(b) Using considerable extra effort, we can relax the restriction $\Lambda > 3$, apparently to $\Lambda > 2$. However, E.A. Rahmanov announced in [10] asymptotics for at least the weights $\exp(-|x|^{\alpha})$, $\alpha > 0$, and before extending the above results, their generality and techniques should be compared to those in forthcoming results of Rahmanov.

(c) P. Nevai [23,pp.140–141] conjectured for $\exp(-|x|^{\alpha})$, $\alpha > 1$, a result like that above, but with $O(n^{-\delta})$ replaced by $O(n^{-1})$. For weights such as $\exp(-x^4)$ or $\exp(-x^6)$, Nevai's conjecture was proved in [1,22,23,29].

(d) One can give a definite numerical value to δ if one carefully follows the proofs.

II. Preliminaries

First, some technical inequalities:

<u>Lemma II.1</u>

Let $W \in FR_\Lambda$, some $\Lambda > 0$. Then

(a)

$$Q'(1)\, x^{\Omega-1} \le Q'(x) \le Q'(1)\, x^{\Lambda-1},\ x \in (0,1]. \tag{II.1}$$

and

$$Q'(1)\, x^{\Lambda-1} \le Q'(x) \le Q'(1)\, x^{\Omega-1},\ x \in [1,\infty). \tag{II.2}$$

(b) Given $t > 1$,

$$t^{\Lambda-1} \le Q'(tx)/Q'(x) \le t^{\Omega-1},\ x \in (0,\infty), \tag{II.3}$$

and for x large enough,

$$\Lambda/2 \le xQ'(x)/Q(x) \le 2\Omega. \tag{II.4}$$

(c) There exists C_1 such that

$$n \le a_n Q'(a_n) \le C_1 n. \tag{II.5}$$

(d) Uniformly for $0 < R \le a_n$, $n \ge 1$, $A_{n,R}$ (defined by (4.12)) satisfies

$$A_{n,R} \sim RQ'(R)/n, \tag{II.6}$$

and uniformly for $a_n/2 \le R \le a_n$, $n \ge 1$,

$$A_{n,R} \sim 1. \tag{II.7}$$

(e) We have

$$\Lambda \le T(x) \le \Omega,\ x \in (0,\infty), \tag{II.8}$$

and

$$\Lambda-1 \le xQ''(x)/Q'(x) \le \Omega - 1, \; x \in (0, \infty). \tag{II.9}$$

(f)

$$1/(\Omega u) \le a_u'/a_u \le 1/(\Lambda u), \; u \in (0, \infty), \tag{II.10}$$

and

$$(s/t)^{1/\Omega} \le a_s/a_t \le (s/t)^{1/\Lambda}, \; 0 \le t \le s < \infty. \tag{II.11}$$

In particular,

$$s^{1/\Omega} \le a_s/a_1 \le s^{1/\Lambda}, \; s \in [1, \infty). \tag{II.12}$$

(g) If also $W \in FR_\Lambda(3)$, then

$$T'(x) = O(1/x), \; x \to \infty. \tag{II.13}$$

<u>Proof</u>

(a) to (d) See Lemma 6.1 in [14,pp.43–44].

(e) In view of (1.13), these are restatements of (I.1).

(f) Firstly, (II.10) follows from the identity (3.23), the inequality (II.8), and the definition (1.10) of a_u. To obtain (II.11), we use

$$a_s/a_t = \exp\left(\int_t^s a_u'/a_u \, du\right),$$

and (II.10).

(g) This follows from (I.2) with j = 3, from (II.9), and the identity

$$T'(x) = \frac{xQ'''(x)}{Q'(x)} + \frac{Q''(x)}{Q'(x)} - x\left(\frac{Q''(x)}{Q'(x)}\right)^2. \qquad \square$$

Next, we recall a Markov–Bernstein inequality. As usual, $Q^{[-1]}$ denotes the inverse function of Q, satisfying $Q^{[-1]}(Q(x)) = x$.

<u>Lemma II.2</u>

Let $W \in FR_\Lambda$ for some $\Lambda > 0$.

(a) There exists $C > 0$ such that

$$\|P'W\|_{L_\infty(\mathbb{R})} \le \{\int_1^{Cn} ds/Q^{[-1]}(s)\} \|PW\|_{L_\infty(\mathbb{R})}, \tag{II.14}$$

for all $n \ge 1$ and $P \in \mathcal{P}_n$.

(b) If in addition, $\Lambda > 1$, then

$$\int_1^{Cn} ds/Q^{[-1]}(s) \le C_2\, n/a_n. \tag{II.15}$$

Proof

(a) This is Theorem 1.1 in [9].

(b) Now

$$\int_1^{Cn} ds/Q^{[-1]}(s) = \int_{Q^{[-1]}(1)}^{Q^{[-1]}(Cn)} Q'(t)/t\, dt$$

$$\le (\Lambda - 1)^{-1} \int_{Q^{[-1]}(1)}^{Q^{[-1]}(Cn)} Q''(t)\, dt \quad \text{(by (II.9))}$$

$$\le (\Lambda - 1)^{-1}\, Q'(Q^{[-1]}(Cn)).$$

Now, from (II.4) and (II.5), we deduce that $Q(a_n) \sim n$, n large enough, and hence for some fixed k, and n large enough,

$$Q^{[-1]}(Cn) \le a_{kn}.$$

Next, (II.9) and the restriction $\Lambda > 1$ ensure that $Q'(t)$ is increasing in $(0, \infty)$. Then (II.5) ensures that

$$Q'(Q^{[-1]}(Cn)) = O(kn/a_{kn}) = O(n/a_n). \quad \square$$

Lemma II.3

Let $W \in FR_\Lambda$, some $\Lambda > 0$. Assume the notation of Lemma 4.1.

(a) There exists $1 < p < 2$ and there exists C_1 such that uniformly for $0 < R \le a_n$, $n \ge 1$,

$$\|\mu_{n,R}\|_{L_p[-1,1]} \le C_1. \tag{II.16}$$

(b) There exists C_2, C_3, C_4, C_5 and ϵ_0 such that uniformly for $0 < \epsilon < \epsilon_0$; $a_n/2 \le R \le a_n$; $n \ge 1$, we have

$$-C_2\,\epsilon^{3/2} + C_3\,\epsilon^{1/2}\,(1 - R/a_n) \le U_{n,R}(1+\epsilon)$$

$$\le -C_4\,\epsilon^{3/2} + C_5\,\epsilon^{1/2}\,(1 - R/a_n). \tag{II.17}$$

Proof

These are part of (c) and (d) of Lemma 6.2 in [14,pp.44–45]. □

Following is an analogue of Lemma 4.6 above.

Lemma II.4

Let $W \in FR_\Lambda$, some $\Lambda > 0$. Let $K > 0$ and let $\{c_n\}_{n=1}^{\infty}$ be given by (I.12). Then there exists n_1 and $L > 0$, and for $n \ge n_1$, polynomials $\hat{P}_n$ of degree at most $n - Kn\Delta_n$, for which

$$n^{-L} \le |\hat{P}_n(x)W(c_nx)| \le n^L,\ x \in [-1,1]. \tag{II.18}$$

Proof

First, let us choose

$$-1 =: \hat{y}_{0n} < \hat{y}_{1n} < \dots < \hat{y}_{n+1,n} := 1,$$

such that

$$\int_{\hat{y}_{jn}}^{\hat{y}_{j+1,n}} \mu_{n,a_n}(t)\,dt = \frac{1}{n+1},\ j = 0, 1, \dots n. \tag{II.19}$$

Lemma 6.3 in [14,p.47] with $R = a_n$ (see also Lemma 6.1 there) ensures that we can find polynomials P_n of degree n, with $\{\hat{y}_{jn}\}_{j=1}^{n}$ as zeros, such that if $\hat{y}_{cn} = \hat{y}_{cn}(x)$ denotes the closest of these zeros to x,

$$C_1\,|x - \hat{y}_{cn}| \le |P_n(x)W(a_nx)| \le C_2\,n^{C_3},\ x \in [-1,1]. \tag{II.20}$$

Further,

$$C_4 \le |P_n(x)W(a_n x)| \exp(-nU_{n,a_n}(x)) \le C_2 n^{C_3}, \quad |x| \in [1,2]. \qquad \text{(II.21)}$$

From (II.19) and Hölder's inequality, if $p^{-1} + q^{-1} = 1$, some p, q > 1,

$$\frac{1}{n+1} \le \|\mu_{n,a_n}\|_{L_p[-1,1]} (\hat{y}_{j+1,n} - \hat{y}_{jn})^{1/q}.$$

In view of Lemma II.3(a), for suitable $p \in (1,2)$, we then obtain

$$\hat{y}_{j+1,n} - \hat{y}_{jn} \ge n^{-2q}, \quad 0 \le j \le n, \; n \ge n_1. \qquad \text{(II.22)}$$

Next, from Lemma II.3(b), and recalling that Δ_n is defined by (I.11), we obtain for $1 \le x \le 1 + b\Delta_n$,

$$nU_{n,a_n}(x) \ge -C_5 \, n\Delta_n^{3/2} = -C_5 \log n.$$

Here, b > 0 is fixed but arbitrary. Similarly, we can derive a matching upper bound. So

$$n^{-C_6} \le |P_n(x)W(a_n x)| \le n^{C_7}, \quad 1 \le |x| \le 1 + b\Delta_n.$$

Letting $\alpha > 2q$ (see (II.22)) and letting

$$\mathscr{S}_n := (-1,1) \cap \{\bigcup_{j=1}^{n} (\hat{y}_{jn} - n^{-\alpha-2}, \hat{y}_{jn} + n^{-\alpha-2})\},$$

we then have from (II.20),

$$n^{-C_8} \le |P_n(x)W(a_n x)| \le n^{C_9}, \quad |x| \le 1 + b\Delta_n \text{ s.t. } x \notin \mathscr{S}_n.$$

Defining

$$P_n^*(x) := P_n(x) + iP_n(x+n^{-\alpha-1}),$$

one obtains, **exactly** as in the proof of Lemma 4.6 above, that

$$n^{-C_{10}} \le |P_n^*(x)W(a_n x)| \le n^{C_{11}}, \quad |x| \le 1 + b\Delta_n. \qquad \text{(II.23)}$$

This proves the lemma in the special case $c_n = a_n$; $K = 0$. To prove it in the general case, one proceeds as in Lemma 4.6, replacing n in (II.23) by $\ell := n - \langle 2Kn\Delta_n \rangle$ and then making the substitutions

$$a_\ell x = c_n u; \quad \hat{P}_n(u) := P_\ell^*(x). \qquad \square$$

III. Approximation with Weighted Polynomials

In this section, we prove the following analogue of Theorem 8.1. The only new aspect concerns the derivatives of the weighted polynomials; for $\hat{J} = 0$, this result is Theorem 11.2 in [14,pp.81–82].

<u>Theorem III.1</u>

Let $W \in FR_\Lambda$, some $\Lambda > 0$. Let $\hat{J}$ be a non–negative integer and let

$$H(x) := 1 + \sum_{n=1}^{\infty} h_{2n} x^{2n}, \; h_{2n} \geq 0, \; n = 1, 2, 3, \ldots , \tag{III.1}$$

be an entire function such that

$$H(x)\, W(x) \sim 1, \; x \in \mathbb{R}, \tag{III.2}$$

and for $j = 1, 2, \ldots \hat{J}$,

$$|H^{(j)}(x)/H(x)| \leq C_1\, Q'(x)^j, \; x \text{ large.} \tag{III.3}$$

(If $\hat{J} = 0$, this assumption is omitted). For example, we may choose $H = G_{Q/2}$. Let h be bounded and analytic in $A := \{z: |z| < 1 \text{ and } |\text{Im } z| < \delta\}$, some $\delta > 0$. Let $\eta \geq 0$, and let

$$g(x) := \begin{cases} (1 - x^2)^\eta\, h(x), & x \in (-1,1), \\ 0 & , x \in \mathbb{R}\backslash(-1,1). \end{cases} \tag{III.4}$$

Let y_{1n} be the largest point of equioscillation of $T_{n\infty}(W,x)W(x)$, and let

$$\xi_n := y_{1n}/a_n, \; n \geq 2. \tag{III.5}$$

Then for $n \geq 2$, there exists $P_n \in \mathcal{P}_{n-1}$ such that for $x \in \mathbb{R}$,

$$|g(x) - P_n(x)/H(a_n x)|$$

$$\leq C_1 \|h\|_{L_\infty[-1,1]} \Delta_n^\eta \log\{1 + \frac{\Delta_n}{||x| - \xi_n|}\}, \tag{III.6}$$

and

$$\|g(x) - P_n(x)/H(a_n x)\|_{L_\infty(\mathbb{R})} \leq C_2 \|h\|_{L_\infty[-1,1]} \Delta_n^\eta \log n. \tag{III.7}$$

If $\hat{J} > 0$, then we have for $j = 1, 2, \ldots \hat{J}$, for $n \geq n_1$, and $|x| \leq \xi_n - 1/n$,

$$|(\tfrac{d}{dx})^j \{g(x) - P_n(x)/H(a_n x)\}|$$

$$\leq C_3 \|h\|_{L_\infty[-1,1]} \Delta_n^\eta \log\{1 + \frac{\Delta_n}{||x| - \xi_n|}\} n^{j(\beta+1)}, \tag{III.8}$$

where

$$\beta := \max\{1, 1/\Lambda\}. \tag{III.9}$$

The constants C_1, C_2 and C_3 are independent of n, x and h.

The proof of Theorem III.1 follows exactly that of Theorem 8.1 above, and is similar to that of Theorem 11.2 in [14,p.81]. First, an analogue of Lemma 8.3.

<u>Lemma III.2</u>

Assume the notation and hypotheses of Theorem III.1 and let Γ_n be as in Definition 8.2. Let $I_n(x)$ be defined by (8.15). There exists n_1 such that if $n \geq n_1$ and $P_n(x)$ is the Lagrange interpolation polynomial of degree at most n–1 to $H(a_n x)g(x)$ at the zeros of $T_{n\infty}(W, a_n x)$, then

$$|g(x) - P_n(x)/H(a_n x)| \leq C_1 \|h\|_{L_\infty[-1,1]} I_n(|x|), \ |x| < \xi_n, \tag{III.10}$$

and

$$|P_n(x)/H(a_n x)| \leq C_2 \|h\|_{L_\infty[-1,1]} I_n(|x|), \ |x| > \xi_n. \tag{III.11}$$

Further, for $j = 1, 2, \ldots \hat{J}$, and for $|x| < \xi_n$,

$$|(\tfrac{d}{dx})^j\{g(x) - P_n(x)/H(a_n x)\}|$$
$$\le C_3 \|h\|_{L_\infty[-1,1]}\, I_n(|x|) n^{j(\beta+1)} \max\{1, (\xi_n - |x|)^{-1} n^{-\beta-1}\}^j. \qquad \text{(III.12)}$$

Here C_1 is independent of n, x and h.

<u>Proof</u>

Firstly, we see that (8.16), (8.17), (8.18) and (8.19) remain valid, noting from (II.8) and (III.2) that

$$H(x)\, W(x) \sim 1 \sim T(x),\ x \in \mathbb{R}. \qquad \text{(III.13)}$$

Next, since $Q^{[-1]}$ is positive and increasing in $[1, \infty)$,

$$\int_1^{Cn} ds/Q^{[-1]}(s) \le C_2 n,$$

and hence Lemma II.2 yields for each fixed $k \ge 1$,

$$\|T_{n\infty}^{(k)}(W,x)W(x)\|_{L_\infty(\mathbb{R})} \le C_3\, n^k\, \|T_{n\infty}(W,x)W(x)\|_{L_\infty(\mathbb{R})}$$
$$= C_3\, n^k\, E_{n\infty}(W). \qquad \text{(III.14)}$$

Next, we can write (as in the proof of Lemma 8.3) for $m \ge 1$ and $x \in \mathbb{R}$,

$$(\tfrac{d}{dx})^m \{\frac{1}{H(a_n x)}\} = a_n^m \sum c\, H^{(\ell_1)}(a_n x) H^{(\ell_2)}(a_n x) \ldots H^{(\ell_m)}(a_n x)/H(a_n x)^{m+1},$$

where the coefficients c are independent of H and x, and the sum is over all m–tuples $(\ell_1, \ell_2, \ldots, \ell_m)$ of non–negative integers with $\ell_1 + \ell_2 + \ldots + \ell_m = m$. Then using (III.2) and (III.3), we have if $1 \ge |x| \ge C_4/a_n$, say,

$$|(\tfrac{d}{dx})^m \{\frac{1}{H(a_n x)}\}| \le C_1\, \{a_n Q'(a_n x)\}^m\, W(a_n x)$$

$$\le C_2\, \{a_n Q'(a_n)\, x^{\Lambda-1}\}^m\, W(a_n x), \quad \text{(by (II.3))}$$

$$\le C_3\, W(a_n x)\, n^m \begin{cases} 1 & , \Lambda \ge 1, \\ a_n^{m(1-\Lambda)} & , \Lambda < 1, \end{cases} \qquad \text{(by (II.5))}$$

$$\le C_4\, W(a_n x)\, n^{m\beta}, \qquad \text{(III.15)}$$

by (II.12) and (III.9). Combining (III.15), (III.14) and Leibniz' formula yields for $|x|$

$< \xi_n$,

$$|(\tfrac{d}{dx})^j\{\frac{T_{n\infty}(W,a_nx)/H(a_nx)}{t-x}\}|$$

$$= |\sum_{k=0}^{j}\sum_{m=0}^{j-k}\binom{j}{k}\binom{j-k}{m}(\tfrac{d}{dx})^k T_{n\infty}(W,a_nx)\,(\tfrac{d}{dx})^m\{1/H(a_nx)\}$$

$$\times\,(\tfrac{d}{dx})^{j-k-m}(t-x)^{-1}|$$

$$\leq C_5\sum_{k=0}^{j}\sum_{m=0}^{j-k}(a_nn)^k\,E_{n\infty}(W)\,n^{m\beta}\,|t-x|^{-(j-k-m)-1}$$

$$\leq C_6\,E_{n\infty}(W)\,|t-x|^{-1}\sum_{\ell=0}^{j}n^{(\beta+1)\ell}\,|t-x|^{-(j-\ell)}$$

$$\leq C_7\,E_{n\infty}(W)\,|t-x|^{-1}\,n^{(\beta+1)j}\,\max\{1,\,n^{-\beta-1}\,(\xi_n-|x|)^{-1}\}^j.$$

Combined with (8.19), this last estimate yields (III.12). The inequalities (III.10) and (III.11) follow directly from (8.16) and (8.17) respectively. □

Next, we present an analogue of Lemma 8.4:

<u>Lemma III.3</u>

Let $W \in FR_\Lambda$, some $\Lambda > 0$.

(a) There exists $\epsilon_0 > 0$ such that for each compact subset K of $G := \{z: \mathrm{Re}(z) \in (-1,1);\ 0 < |\mathrm{Im}\, z| < \epsilon_0\}$, we have

$$\limsup_{n\to\infty}\{E_{n\infty}(W)/\min_{z\in K}|T_{n\infty}(W,a_nz)W(a_n|z|)|\}^{1/n} < 1. \tag{III.16}$$

b) Let ℓ be a fixed positive integer. There exists n_1 such that for $n \geq n_1$,

$$z = \xi_ne^{iy},\ 0 \leq y < \pi/2, \tag{III.17}$$

we have

$$E_{n\infty}(W)/|T_{n\infty}(W,a_nz)W(a_n|z|)| \leq (1 + C_1\,(y/\Delta_n)^2\,)^{-\ell}. \tag{III.18}$$

Proof

(a) This is a special case of (10.10) in Corollary 10.3 in [14,p.74].

(b) For a slightly different range of z, this is (10.13) in [14,p.75]. We provide the proof by proceeding as in Lemma 8.4(b): For z given by (III.17),

$$E_{n\infty}(W)/|T_{n\infty}(W,a_n z)W(a_n|z|)|$$

$$= |T_{n\infty}(W,a_n\xi_n)W(a_n\xi_n)|/|T_{n\infty}(W,a_n z)W(a_n\xi_n)|$$

$$= \prod_{x_{jn}^{(\infty)}>0} \left|\frac{(a_n\xi_n)^2 - (x_{jn}^{(\infty)})^2}{(a_n z)^2 - (x_{jn}^{(\infty)})^2}\right|$$

$$= \prod_{x_{jn}^{(\infty)}>0} \left|\frac{\xi_n^2 - z_{jn}^2}{z^2 - z_{jn}^2}\right|,$$

where $z_{jn} := x_{jn}^{(\infty)}/a_n$, $n \geq 1$. Here

$$|z^2 - z_{jn}^2|^2 = (\xi_n^2 - z_{jn}^2)^2 + 4\xi_n^2 z_{jn}^2 \sin^2 y$$

$$\geq (\xi_n^2 - z_{jn}^2)^2 + 16\pi^{-2} \xi_n^2 z_{jn}^2 y^2$$

$$\geq (\xi_n^2 - z_{jn}^2)^2 + C_1 y^2,$$

if $n \geq n_1$ and j is fixed. Finally, for each fixed j,

$$|\xi_n^2 - z_{jn}^2| \leq C_2 \Delta_n,$$

by Theorem 8.1(a) and Corollary 8.2 in [14,p.58] and hence for each fixed ℓ and for $n \geq n_1(\ell)$,

$$E_{n\infty}(W)/|T_{n\infty}(W,a_n z)W(a_n|z|)|$$

$$\leq \prod_{j=1}^{2\ell} [1 + C_1 y^2 (\xi_n^2 - z_{jn}^2)^{-2}]^{-1/2} \prod_{j=2\ell+1}^{n/2-1} 1$$

$$\leq [1 + C_2 y^2/\Delta_n^2]^{-\ell}. \quad \square$$

Following the proof of Theorem 8.1 above, we proceed to the

Proof of Theorem III.1

Proceeding exactly as in the proof of Theorem 8.1, we obtain the estimate (8.25), namely,

$$I_n(|x|) \leq C_{11}\, \Delta_n^{\eta} \log\{1 + \Delta_n/|\xi_n - |x||\}, \tag{III.19}$$

for $x \in \mathbb{R}$ and $n \geq n_1$. The only difference is that instead of Lemma 8.4, one uses Lemma III.3, and instead of Corollary 7.2 above, one uses Corollary 8.2 in [14,p.58].

Armed with (III.19), we immediately obtain (III.6) from (III.10), at least for $|x| < \xi_n$, and for $|x| > 1$. The range $\{x: |x| \in [\xi_n,1]\}$ can be filled exactly as in the proof of Theorem 8.1, or the proof of Theorem 11.2 in [14]. The derivative estimate (III.8) follows from (III.12) and (III.19), and the estimate (III.7) may be completed as in the proof of Theorem 8.1. The fact that we may choose $H = G_{Q/2}$ follows from Theorem 6.1(a). □

IV. One-Sided Weighted Polynomial Approximation

In this section, we prove the following analogue of Theorem 9.1:

<u>Theorem IV.1</u>

Let $W \in FR_\Lambda$, some $\Lambda > 0$. Let H be as in Theorem III.1, with $\hat{J} = 2$. Fix $s \geq 0$, and define $\{c_n\}_{n=1}^{\infty}$ by (I.12). Further, let $\{k_n\}_{n=1}^{\infty}$ be a sequence of positive integers such that

$$k_n = O(n\Delta_n),\ n \to \infty. \tag{IV.1}$$

Let $\kappa \in [-1,1]$. Then there exists $\delta > 0$ and n_1 such that for $n \geq n_1$ and $|x_0| \leq 1 - n^{-\delta}$, there exist polynomials $P_n(x)$ with real coefficients, of degree at most $2(n - k_n)$, positive in $[-1,1]$ except possibly for simple zeros at ± 1, and with the following properties: Let

$$\tau_n(u) := P_n(u)/H^2(c_n u)\ (1 - u^2)^{-\kappa},\ u \in (-1,1). \tag{IV.2}$$

Then we can ensure that either

$$\tau_n(u) \leq 1,\ u \in (-1,1) \text{ if } \kappa \in [-1,1), \tag{IV.3}$$

or

$$\tau_n(u) \geq 1,\ u \in (-1,1), \text{ if } \kappa \in [0,1]. \tag{IV.4}$$

Further,

$$\tau_n(x_0) = 1;\ \tau_n'(x_0) = 0; \tag{IV.5}$$

and for $j = 0, 1, 2,$

$$\left|\left(\frac{d}{du}\right)^j \{\tau_n(u) - 1\}\right| \leq n^{-\delta},\ |u| \leq 1 - n^{-\delta}. \tag{IV.6}$$

Finally

$$\int_{-1}^{1} |\log \tau_n(t)| \ (1-t^2)^{-1/2} \, dt \le n^{-\delta}, \tag{IV.7}$$

and

$$(1-u^2)\, n^{-1/\delta} \le \tau_n(u) \le (1-u^2)^{-|\kappa|}\, n^{1/\delta},\ u \in [-1,1]. \tag{IV.8}$$

For the P_n satisfying (IV.4), or satisfying (IV.3) with $\kappa < 0$, we may in addition ensure that P_n is positive in $[-1,1]$.

We note that Lemmas 9.2, 9.3, and 9.4 did not involve W, and so we can apply them in the present situation. Following is an analogue of Lemma 9.5.

<u>Lemma IV.2</u>

Assume the notation and hypotheses of Theorem IV.1. There exist L and n_1 such that for $n \ge n_1$, there exists a polynomial $\tilde{P}_n$ of degree at most $n - k_n$ such that

$$n^{-L} \le |\tilde{P}_n(x)/H(c_n x)| \le n^{-2},\ x \in [-1,1], \tag{IV.9}$$

and for j = 1, 2,

$$\left|\left(\frac{d}{dx}\right)^j \{\tilde{P}_n(x)/H(c_n x)\}\right| \le n^{-2},\ x \in [-1,1]. \tag{IV.10}$$

<u>Proof</u>

Lemma II.4 provides polynomials $\hat{P}_n$ of degree at most $n - k_n$ (recall that k_n satisfies (IV.1)) for which (II.18) holds. Applying the Markov–Bernstein inequality in Lemma II.2 (after a suitable transformation) yields for $x \in [-1,1]$, and j = 1, 2,

$$|\hat{P}_n^{(j)}(x) W(c_n x)| \le n^L \left\{c_n \int_1^{Cn} ds/Q^{[-1]}(s)\ ds\right\}^j$$

$$\le C_1\, n^L\, (n^{1+1/\Lambda})^j. \tag{IV.11}$$

Next, a minor modification of (III.15) ensures that for $x \in [-1,1]$, and fixed $m \ge 1$,

$$\left|\left(\frac{d}{dx}\right)^m \{1/H(c_n x)\}\right| \le C_2\, W(c_n x)\, n^{m\beta}. \tag{IV.12}$$

Applying (IV.11), (IV.12) and Leibniz' formula leads to an estimate of the form

$$|(\tfrac{d}{dx})^j \{\hat{P}_n(x)/H(c_n x)\}| \le n^{C_3}, \; x \in [-1,1], \; j = 1, 2.$$

Defining

$$\tilde{P}_n(x) := n^{-b} \hat{P}_n(x),$$

with b large enough, yields (IV.10), while (II.18) yields (IV.9). □

Proof of Theorem IV.1

This is the same as that of Theorem 9.1, involving the same seven steps, except for some minor changes. Instead of Theorem 8.1, we use Theorem III.1; and instead of Lemma 9.5, we use Lemma IV.2; and instead of approximating $(1 - x^2)^6$ in Step 1, we approximate $(1 - x^2)^\lambda$ with λ large enough. Following are the details:

Step 1 Approximate $(1 - x^2)^\lambda$

Let λ be some fixed, but large enough, positive number, and let

$$g(x) := \begin{cases} (1 - x^2)^\lambda, & x \in [-1,1], \\ 0 & , \text{otherwise}. \end{cases}$$

By (III.7) and (III.8), we can find polynomials $P_n(x)$ of degree at most n such that for $n \ge n_1$,

$$\|g(x) - P_n(x)/H(a_n x)\|_{L_\infty(\mathbb{R})} \le C_1 \Delta_n^\lambda \log n,$$

and for $|x| \le 1 - C_2 \Delta_n$, and j = 1, 2,

$$|(\tfrac{d}{dx})^j\{g(x) - P_n(x)/H(a_n x)\}| \le C_3 \Delta_n^\lambda n^{j(\beta+1)}.$$

Since β (given by (III.9)) depends only on Λ, we can choose $\lambda \ge 6$ so large that for $n \ge n_1$,

$$\|g(x) - P_n(x)/H(a_n x)\|_{L_\infty(\mathbb{R})} \le n^{-2}, \tag{IV.13}$$

and for j = 1, 2, and $|x| \le 1 - C_2 \Delta_n$,

$$|(\tfrac{d}{dx})^j\{g(x) - P_n(x)/H(a_n x)\}| \le n^{-2}. \tag{IV.14}$$

Step 2 From a_n to c_n and degree n to degree $n - <\alpha n\Delta_n>$

One replaces n in (IV.13) and (IV.14) by $\ell := n - <\alpha n\Delta_n>$, where α is a fixed but large enough positive number. Then one makes the substitutions

$$a_\ell x = c_n u; \quad P_\ell(x) =: P_n^*(u).$$

Exactly as in the proof of Theorem 9.1, this leads to

$$\|g(u) - P_n^*(u)/H(c_n u)\|_{L_\infty(\mathbb{R})} \le C_8 \Delta_n, \tag{IV.15}$$

and

$$|(\tfrac{d}{du})^j \{g(u) - P_n^*(u)/H(c_n u)\}| \le C_8 \Delta_n, \tag{IV.16}$$

for $j = 1, 2$ and $|u| \le 1 - C_9 \Delta_n$. Note too that P_n^* has degree at most $n - <\alpha n\Delta_n>$.

Step 3: Approximate 1.

In the subsequent analysis, it will be easier to work with approximations to 1 rather than to $(1 - x^2)^\lambda$. For this purpose, we use $S_m^*(u)$ of Lemma 9.3, with $\tau = \lambda$, $x_0 = 0$, and with $m := <n^{1/(2+2\lambda)}> + 1$. Let

$$R_n(u) := P_n^*(u)S_m^*(u),$$

a polynomial of degree at most $n - <\frac{\alpha}{2} n\Delta_n>$, n large enough. (Recall that $\lambda \ge 6$).

Firstly, (9.21) and (IV.15), (IV.16) yield (as in Step 3 of the proof of Theorem 9.1),

$$|(\tfrac{d}{du})^j \{R_n(u)/H(c_n u) - 1\}| \le C_{10}\, n^{-1/2},\ j = 0, 1, 2, \tag{IV.17}$$

$|u| \le 1 - n^{-1/(9\lambda)}$ (recall that $\lambda \ge 6$). Further, from (9.19), (9.20), and (IV.15),

$$\begin{aligned} |R_n(u)/H(c_n u)| &\le S_m^*(u)\{g(u) + C_{11} \Delta_n\} \\ &\le 1 + C_{11}\, m^{\lambda+1} \{\tfrac{\log n}{n}\}^{2/3} \\ &\le 1 + C_{12}\, n^{-1/7},\ u \in [-1,1]. \end{aligned} \tag{IV.18}$$

Step 4: Interpolation at x_0.

For a given large enough n, let $f(u) := R_n(u)/H(c_n u)$, $u \in [-1,1]$, and for a given x_0, define a linear polynomial

$$F(u) := F(u;x_0) := 1/f(x_0) - (u - x_0)\,(f'/f^2)(x_0).$$

Then defining

$$R_n^*(u) := R_n(u)F(u),$$

a polynomial of degree at most $n - <\frac{\alpha}{2}\, n\Delta_n> + 1$, we have

$$\left(\frac{d}{dx}\right)^j\{R_n^*(x)/H(c_n x) - 1\}_{|x = x_0} = 0,\ j = 0,\ 1, \tag{IV.19}$$

and for $j = 0, 1, 2$,

$$\left|\left(\frac{d}{dx}\right)^j\{R_n^*(x)/H(c_n x) - 1\}\right| \le C_{13}\, n^{-1/2}, \tag{IV.20}$$

for the range

$$|x|\,;\ |x_0| \le 1 - n^{-1/(9\lambda)}, \tag{IV.21}$$

and

$$|R_n^*(x)/H(c_n x)| \le 1 + C_{14} n^{-1/7}, \tag{IV.22}$$

for the range

$$|x| \le 1;\ |x_0| \le 1 - n^{-1/(9\lambda)}. \tag{IV.23}$$

Step 5: Beef up the weighted polynomials near ± 1.

We use the polynomials $\tilde{P}_n(x)$ of Lemma IV.2 of degree at most $n - <\frac{\alpha}{4} n\Delta_n>$, n large enough. Define

$$Q_n(x) := R_n^*(x)^2 + \tilde{P}_n(x)^2\,(x - x_0)^2,$$

a polynomial of degree at most $2n - 2<\frac{\alpha}{4} n\Delta_n>$, n large enough. From (IV.20) and (IV.9), (IV.10), for the range (IV.21),

$$\left|\left(\frac{d}{dx}\right)^j\{Q_n(x)/H^2(c_n x) - 1\}\right| \le C_{15}\, n^{-1/2},\ j = 0,\ 1,\ 2. \tag{IV.24}$$

Further, from (IV.19),

$$(\frac{d}{dx})^j\{Q_n(x)/H^2(c_nx) - 1\}_{|x=x_0} = 0,\ j = 0, 1. \tag{IV.25}$$

Suppose now that $n \geq n_1$ and $|x_0| \leq 1 - 2n^{-1/(9\lambda)}$. Then for $|x| \leq 1 - n^{-1/(9\lambda)}$, (IV.20) ensures that

$$Q_n(x)/H^2(c_nx) \geq \{R_n^*(x)/H(c_nx)\}^2 \geq 1 - C_{13}\, n^{-1/2} \geq 1/2,$$

while if $1 \geq |x| \geq 1 - n^{-1/(9\lambda)}$, then (IV.9) shows that for some $L_1 > 0$,

$$Q_n(x)/H^2(c_nx) \geq n^{-2/(9\lambda)}\, \tilde{P}_n(x)^2/H^2(c_nx) \geq n^{-L_1}.$$

Hence, using (IV.22) and (IV.9), we obtain,

$$n^{-L_1} \leq Q_n(x)/H^2(c_nx) \leq 1 + C_{16}\, n^{-1/7}, \tag{IV.26}$$

for $x \in [-1,1]$, $|x_0| \leq 1 - 2n^{-1/(9\lambda)}$.

Step 6 Polynomials satisfying (IV.3).

By expanding $Q_n(x)/H^2(c_nx)$ about $x = x_0$, and using (IV.24), (IV.25), we obtain for some ξ between x and x_0,

$$Q_n(x)/H^2(c_nx) = 1 + \frac{1}{2}(x - x_0)^2 (\frac{d}{dx})^2\{Q_n(x)/H^2(c_nx)\}_{|x=\xi}$$

$$\leq 1 + C_{17}\, n^{-1/2}(x - x_0)^2, \tag{IV.27}$$

$|x| \leq 1 - n^{-1/(9\lambda)}$, $|x_0| \leq 1 - 2n^{-1/(9\lambda)}$. Let us suppose first $\kappa < 0$ and let $m := \langle n^{|\kappa|/(20\lambda)}\rangle$, and choose $S_m^*(x)$ to be the polynomial of Lemma 9.3 with $h(u) := (1 - u^2)^\kappa$, so that there $\tau = -\kappa$. Define for some fixed $b > 0$,

$$P_n(x) := Q_n(x)\{S_m^*(x) - n^{-b}(x - x_0)^2\},$$

a polynomial of degree $\leq 2n - \langle\frac{\alpha}{8}n\Delta_n\rangle$, n large enough. (Recall that $|\kappa| \leq 1$ and $\lambda \geq 6$). If α is large enough, it follows that $P_n(x)$ has degree at most $2n - 2k_n$, as required. Note first from (IV.25) and from (9.18), that

$$\tau_n(x) := \{P_n(x)/H^2(c_nx)\}(1 - x^2)^{-\kappa},$$

satisfies the interpolation conditions (IV.5). Next, from (9.19), and (IV.26), $P_n(x)$ is positive in $[-1,1]$. Next, from (IV.24) and (9.21), we see that (IV.6) holds uniformly for $|u| \le 1 - n^{-\delta}$, $|x_0| \le 1 - n^{-\delta}$, δ small enough. Further, from (IV.26) and (9.19), we obtain something better than the upper and lower bounds in (IV.8). Then (IV.7) follows from (IV.8) and (IV.6), possibly with a smaller δ. It remains to establish (IV.3), and this is the hard part. First, from (IV.27) and (9.19),

$$\begin{aligned}\tau_n(x) &= \{Q_n(x)/H^2(c_n x)\}\,\{S_m^*(x)(1-x^2)^{-\kappa} - n^{-b}(x-x_0)^2\,(1-x^2)^{-\kappa}\}\\ &\le \{1 + C_{17}\, n^{-1/2}(x-x_0)^2\}\{1 - n^{-b}(x-x_0)^2(1-x^2)^{-\kappa}\}\\ &\le 1,\end{aligned}$$

for $|x| \le 1 - n^{-1/(9\lambda)}$, $|x_0| \le 1 - 2n^{-1/(9\lambda)}$, provided for example,

$$C_{17}\, n^{-1/2} < n^{-b}\,(1-x^2)^{-\kappa},$$

which will be true (recall that $\kappa < 0$) if

$$b < 1/2 - |\kappa|/(9\lambda). \tag{IV.28}$$

Also for $1 \ge |x| \ge 1 - n^{-1/(9\lambda)}$, we obtain from (IV.26) and (9.20) that

$$\begin{aligned}\tau_n(x) &\le 2\, m^{1-\kappa}\,(1-x^2)^{-\kappa}\\ &\le 2\, C_{18}\, n^{(1-\kappa)|\kappa|/(20\lambda) - |\kappa|/(9\lambda)} \to 0,\end{aligned}$$

$n \to \infty$, since $(1-\kappa) \le 2$ and by our choice $m = \langle n^{|\kappa|/(20\lambda)}\rangle$. Hence (IV.3).

In the case $1 > \kappa \ge 0$, we take the above polynomials for $(1-x^2)^{\kappa-1}$, and then absorb the extra factor $1 - x^2$ in τ into the polynomial P_n. Thus we write

$$\{P_n(u)/H^2(c_n u)\}(1-u^2)^{-(\kappa-1)} =: \{P_n^{\#}(u)/H^2(c_n u)\}(1-u^2)^{-\kappa}.$$

It is not difficult to see that all of (IV.3), (IV.5–8) are preserved. For the lower bound in (IV.8), one needs to use that the polynomials for $(1 - x^2)^{\kappa-1}$ admit a bound like (IV.26), and to recall the lower bound for S_m^* in (9.19).

Step 7: Polynomials satisfying (IV.4).

This is more complicated than Step 6, as the polynomials of Step 5 may be too small near ± 1. We start with the polynomials $Q_n(x)$ satisfying (IV.24–26), and multiply by the polynomials $R_m(x)$ of Lemma 9.2, with $m := <n^{1/(2\lambda)}>$; $\sigma := 1$; K fixed but large enough, and $\epsilon_m := n^{-1/(9\lambda)}$. More precisely, set

$$U_n(x) := Q_n(x) \{R_m(x) + (1 - R_m(x_0)) + \frac{d}{dt}(1 - R_m(t))|_{t=x_0}(x - x_0)\},$$

a polynomial of degree at most $2n - 2<\frac{\alpha}{8} n\Delta_n>$, n large enough. We have from (IV.25),

$$(\frac{d}{dx})\{U_n(x)/H^2(c_n x) - 1\}|_{x=x_0} = 0, \ j = 0, 1. \tag{IV.29}$$

Further, from (IV.24), and (9.11), (9.12),

$$|(\frac{d}{dx})^j\{U_n(x)/H^2(c_n x) - 1\}| \leq C_{19} n^{-1/(10\lambda)}, \tag{IV.30}$$

for j = 0, 1, 2, and for

$$|x_0| \leq 1 - 2n^{-1/(9\lambda)}, \ |x| \leq 1 - n^{-1/(9\lambda)} - m^{-1/4}, \tag{IV.31}$$

since $m^{-1/4} \leq n^{-1/(9\lambda)}$, n large enough. Next, from (9.10) and (IV.26), if K in Lemma 9.2 is large enough, and $1 \geq |x| \geq 1-\epsilon_m = 1 - n^{-1/(9\lambda)}$,

$$U_n(x)/H^2(c_n x) \geq n^{-L_1}(m^K - C_{20} m^{-1/5}) \geq 1.$$

Similarly for an upper bound, so that for some fixed S, independent of x, x_0, and n,

$$n^S \geq U_n(x)/H^2(c_n x) \geq 1, \ 1 - n^{-1/(9\lambda)} \leq |x| \leq 1. \tag{IV.32}$$

Also from (9.13) and (IV.24),

$$\left.\begin{aligned} &n^S \geq U_n(x)/H^2(c_n x) \geq 1 - C_{21} n^{-1/\lambda}, \\ &1 - n^{-1/(9\lambda)} \geq |x| \geq 1 - n^{-1/(9\lambda)} - m^{-1/4}. \end{aligned}\right\} \tag{IV.33}$$

Finally, we apply (as in Theorem 9.1), the polynomials S_m^* of Lemma 9.4 for $h(u) := (1 - u^2)^\kappa$, (so that there $\tau = \kappa$) with m = n. More precisely, let $\epsilon \in (0,1)$ be fixed, and let

$$P_n(x) := U_n(x)\ \{S_m^*(x) + n^{-\epsilon}(x - x_0)^2\}.$$

For τ_n given by (IV.2), both (IV.5) and (IV.6) follow from (9.18), (9.21), (IV.29) and (IV.30). Further, (IV.8) (and hence (IV.7)) follows from (IV.30), (IV.32), (IV.33) and (9.23). It remains to prove (IV.4). First, by (IV.29), (IV.30) and (9.23), for $|x| \le 1 - n^{-1/(9\lambda)} - m^{-1/4}$ and $n \ge n_1$,

$$\tau_n(x) = \{U_n(x)/H^2(c_nx)\}\{S_m^*(x)(1-x^2)^{-\kappa} + n^{-\epsilon}(x-x_0)^2(1-x^2)^{-\kappa}\},$$

$$\ge \{1 - C_{22}n^{-1/(10\lambda)}(x-x_0)^2\}\{1 + n^{-\epsilon}(x-x_0)^2(1-x^2)^{-\kappa}\}$$

$$= 1 + n^{-\epsilon}(x-x_0)^2\{(1-x^2)^{-\kappa} - C_{22}n^{-1/(10\lambda)+\epsilon}$$

$$- C_{22}n^{-1/(10\lambda)}(x-x_0)^2(1-x^2)^{-\kappa}\}$$

$$\ge 1 + n^{-\epsilon}(x - x_0)^2\{\tfrac{1}{2}\ (1 - x^2)^{-\kappa} - C_{22}n^{-1/(10\lambda)+\epsilon}\} \ge 1,$$

if $|x| \le 1 - n^{-1/(9\lambda)} - m^{-1/4}$, and $\epsilon < 1/(10\lambda)$. (Recall that $\kappa \in [0,1]$). Thus, we have (IV.4) for such x. For $1 \ge |x| \ge 1 - n^{-1/(9\lambda)} - m^{-1/4}$, we have from (IV.32), (IV.33), and (9.23), provided $|x_0| \le 1 - 3n^{-1/(9\lambda)}$,

$$\tau_n(x) \ge \{1 - C_{23}n^{-1/\lambda}\}\{1 + n^{-\epsilon}(x - x_0)^2\ (1 - x^2)^{-\kappa}\}$$

$$\ge \{1 - C_{23}n^{-1/\lambda}\}\{1 + \tfrac{1}{2}\ n^{-\epsilon-2/(9\lambda)}\} \ge 1,$$

provided also $\epsilon < 1/\lambda - 2/(9\lambda) = 7/(9\lambda)$. This completes the proof of (IV.4). □

V. Asymptotics of Orthogonal Polynomials on IR

In this section, we prove analogues for Freud weights of the results in Section 13, and in particular prove Theorem I.4. First, an analogue of Theorem 13.1:

Theorem V.1

Let $W \in FR_\Lambda$, some $\Lambda > 0$. Let H be as in Theorem III.1, with $\hat{J} = 2$. Fix $s \geq 0$, and define $\{c_n\}_{n=1}^{\infty}$ by (I.12). Let $\{k_n\}_{n=1}^{\infty}$ be a sequence of positive integers satisfying (IV.1). For n = 1, 2, 3, ..., let $S_n(x)$ be a polynomial positive in $[-1,1]$ and of degree at most $2k_n$. Let $0 < \beta < 2$ and for $n \geq 1$, let

$$f_n(\theta) := |\sin\,\theta|^{\beta}/\{H^2(c_n \cos\,\theta)S_n(\cos\,\theta)\},\ \theta \in [-\pi,\pi]. \tag{V.1}$$

Let $L > 0$. Then there exist n_1 and $\delta > 0$ such that uniformly for $n \geq n_1$, $\phi \in [n^{-\delta},\pi - n^{-\delta}]$, and

$$m \geq 2n - 2Ln\Delta_n, \tag{V.2}$$

we have

$$|\phi_m(f_n,z)/\{z^m/\overline{D(f_n;z)}\} - 1| \leq n^{-\delta},\ z := e^{i\phi}. \tag{V.3}$$

Proof

This is the same as that of Theorem 13.1, but we provide the details. We shall use Lemma 11.6 and the polynomials of Theorem IV.1, with $\kappa = -\beta/2$ and $\kappa = (2-\beta)/2$. By Theorem IV.1, there exist for n large enough, polynomials $P_n^+(u)$, $P_n^-(u)$ of degree at most $2n - 2k_n - 2Ln\Delta_n - 2$, which are positive in $[-1,1]$ (note that we use (IV.3) with $\kappa < 0$), and satisfy

$$(1-u^2)^{(2-\beta)/2}/P_n^-(u) \leq 1/H^2(c_n u) \leq (1-u^2)^{-\beta/2}/P_n^+(u),\ u \in (-1,1).$$

Further, (IV.5) to (IV.8) are valid. Then defining

$$f_n^+(\theta) := 1/\{P_n^+(\cos\theta)S_n(\cos\theta)\},\ \theta\in[-\pi,\pi],$$

and

$$f_n^-(\theta) := (\sin\theta)^2/\{P_n^-(\cos\theta)S_n(\cos\theta)\},\ \theta\in[-\pi,\pi],$$

we have from (V.1),

$$f_n^-(\theta) \le f_n(\theta) \le f_n^+(\theta),\ \theta\in[-\pi,\pi]. \tag{V.4}$$

The degree of the trigonometric polynomial $P_n^{\pm}(\cos\theta)S_n(\cos\theta)$ is at most $2n - 2Ln\Delta_n - 2$. Next,

$$D(f_n^+;0)/D(f_n^-;0) = \exp\left(\frac{1}{2\pi}\int_0^\pi \log\,[f_n^+(\theta)/f_n^-(\theta)]\,d\theta\right)$$

$$= \exp\left(\frac{1}{2\pi}\int_0^\pi \log\,[f_n^+(\theta)/f_n(\theta)]\,d\theta + \frac{1}{2\pi}\int_0^\pi \log\,[f_n(\theta)/f_n^-(\theta)]\,d\theta\right).$$

Here, with an obvious modification of the notation in Theorem IV.1,

$$f_n(\theta)/f_n^+(\theta) = \{P_n^+(\cos\theta)/H^2(c_n\cos\theta)\}\,|\sin\theta|^\beta =: \tau_n^+(\cos\theta),$$

and

$$f_n(\theta)/f_n^-(\theta) = \{P_n^-(\cos\theta)/H^2(c_n\cos\theta)\}\,|\sin\theta|^{-(2-\beta)} =: \tau_n^-(\cos\theta).$$

According to (IV.7) for τ_n^+ and τ_n^-, we then have

$$D(f_n^+;0)/D(f_n^-;0) = \exp(O(n^{-\delta})) = 1 + O(n^{-\delta}),\ n\to\infty. \tag{V.5}$$

Next, from (IV.5), given $x_0 = \cos\phi$, where for some $\delta_1 > 0$, $\phi\in[n^{-\delta_1},\pi-n^{-\delta_1}]$,

$$f_n^-(\phi) = f_n(\phi) = f_n^+(\phi). \tag{V.6}$$

Further,

$$I := \int_{-\pi}^{\pi}\frac{\log\{f_n^+(t)/f_n^-(t)\}}{(\sin(\frac{t-\phi}{2}))^2}\,dt$$

$$= \int_0^\pi \log\{\tau_n^-(\cos t)/\tau_n^+(\cos t)\}\,\{[\sin(\tfrac{t-\phi}{2})]^{-2}+[\sin(\tfrac{t+\phi}{2})]^{-2}\}\,dt$$

$$\le \pi^2\int_0^\pi\frac{\log\{\tau_n^-(\cos t)/\tau_n^+(\cos t)\}}{(t-\phi)^2}\,dt + C_1 n^{2\delta_1}\int_0^\pi \log\{\tau_n^-(\cos t)/\tau_n^+(\cos t)\}\,dt$$

(as $|\sin(\frac{t-\phi}{2})| \geq \pi^{-1}|t-\phi|$ and $|\sin(\frac{t+\phi}{2})| \geq C_2 n^{-\delta_1}$)

$$\leq \pi^2 n^{\delta/2} \int_{|t-\phi|\geq n^{-\delta/4}} \log\{\tau_n^-(\cos t)/\tau_n^+(\cos t)\}\, dt$$

$$+ \pi^2 \int_{|t-\phi|\leq n^{-\delta/4}} \tfrac{1}{2}\left|\left(\tfrac{d}{dt}\right)^2[\log \tau_n^-(\cos t)-\log \tau_n^+(\cos t)]_{|t=\xi}\right| dt$$

$$+ C_1 n^{2\delta_1} \int_0^{\pi} \log\{\tau_n^-(\cos t)/\tau_n^+(\cos t)\}\, dt.$$

In the second integral, we have expanded $\log \tau_n^-(\cos t) - \log \tau_n^+(\cos t)$ about $t = \phi$ in a Taylor series to second derivatives, and have used (IV.5). The number $\xi = \xi(n,t,\phi)$ lies between t and ϕ. Using (IV.6) and (IV.7), we obtain if δ_1 is small enough,

$$I \leq C\, n^{-\delta/2} + C\, n^{-\delta/4-\delta} + C\, n^{2\delta_1-\delta} \leq C_3\, n^{-\delta/2}.$$

Then (V.5) and Lemma 11.6 show that if δ is small enough, then uniformly for $n \geq n_1$, $\phi \in [n^{-\delta}, \pi-n^{-\delta}]$, $z := e^{i\phi}$ and m satisfying (V.2),

$$|\phi_m(f_n,z)/\{z^m/\overline{D(f_n^+;z)}\} - 1| \leq C\, n^{-\delta/2}. \tag{V.7}$$

It remains to show that for $n \geq n_1$ and uniformly for the relevant range of z,

$$|D(f_n^+;z)/D(f_n;z) - 1| = |D(f_n^+/f_n;z) - 1| \leq n^{-\delta}. \tag{V.8}$$

According to (11.24) in Lemma 11.3(b), since (V.6) holds, this is equivalent to estimation of

$$J := \int_{-1}^{1} \left|\frac{\log \tau_n^+(s) - \log \tau_n^+(x)}{s - x}\right| (1-s^2)^{-1/2}\, ds,\ x = \cos \phi.$$

This may be estimated using (IV.6) and (IV.7) in a somewhat easier fashion than I, and leads to (V.8). □

To prove an analogue of Theorem 13.2, we need an analogue of Lemma 10.2. We remark that it is only in this lemma that we need $\Lambda > 3$, in approximating via

Jackson's Theorem on an interval of length $c_n = O(n^{1/\Lambda})$, by polynomials of degree $\leq n^{1/3}$.

Lemma V.2

Let $W \in FR_\Lambda(3)$, some $\Lambda > 3$. Let $G_{Q/2}$ be the associated entire function, and let (as in (10.9)),

$$\Psi(x) := 1/\{G_{Q/2}W\}(x),\ x \in \mathbb{R}. \tag{V.9}$$

Let $\{c_n\}_{n=1}^{\infty}$ be given by (I.12) with $s \geq 0$ fixed. Then there exists $\lambda > 0$ and for n large enough, polynomials R_n of degree at most $n^{1/3}$ such that

$$|R_n(x)/\Psi(c_nx) - 1| \leq C_1\, n^{-\lambda},\ x \in [-2,2], \tag{V.10}$$

and

$$|(\tfrac{d}{dx})\{R_n(x)/\Psi(c_nx)\}| \leq C_2\, n^{1/2},\ x \in [-1,1]. \tag{V.11}$$

Proof

From Theorem 6.1(b) and (II.8), we have for x large enough,

$$\Psi(x) = \{\pi\, T(x)\}^{-1/2} + O(Q(x)^{-1/2}), \tag{V.12}$$

and

$$\Psi'(x) = \Psi(x)\ \{Q'(x) - G'_{Q/2}(x)/G_{Q/2}(x)\} \tag{V.13}$$

$$= O(Q'(x)/Q(x)^{1/2}),$$

$$= O(\{Q'(x)/x\}^{1/2}), \tag{V.14}$$

by (II.4). Recall too from (II.13) that

$$T'(x) = O(1/x),\ x \to \infty. \tag{V.15}$$

We use these estimates on the modulus of continuity

$$\omega_n(\eta) := \max_{\substack{|x-y| \leq \eta \\ x,\, y \in \mathbb{R}}} |\Psi(c_nx) - \Psi(c_ny)|,\ \eta > 0.$$

Fix some small $\rho > 0$, and first let

$$n^\rho/c_n \leq x \leq y \leq x + \eta.$$

Then from (V.12) and (V.15), for some ξ between x and y,

$$
\begin{aligned}
&|\Psi(c_n x) - \Psi(c_n y)| \\
&= |\{\pi\, T(c_n x)\}^{-1/2} - \{\pi\, T(c_n y)\}^{-1/2} + O(Q(c_n x)^{-1/2})| \\
&= O(c_n |x - y| / (c_n x)) + O(Q(n^\rho)^{-1/2}) \\
&= O(c_n\, \eta\, n^{-\rho}) + O(n^{-\Lambda\rho/2}),
\end{aligned}
$$

by (II.2) and (II.4), which shows that

$$Q(x) \geq C_3\, x^\Lambda, \text{ x large enough.}$$

Finally, using the estimate (II.12), we obtain

$$|\Psi(c_n x) - \Psi(c_n y)| \leq C_4\, [n^{1/\Lambda-\rho}\, \eta + n^{-\Lambda\rho/2}], \tag{V.16}$$

$n^\rho/c_n \leq x \leq y \leq x + \eta$.

Next, from (II.1), it follows that $Q'(x) \to Q'(0) = 0$ as $x \to 0$. Further,

$$
\begin{aligned}
\frac{d}{dt}\{Q'(t)/t\} &= t^{-2}\{tQ''(t) - Q'(t)\} \\
&\geq t^{-2}\, (\Lambda-2)\, Q'(t) > 0,
\end{aligned}
$$

by (II.9), so $Q'(t)/t$ is increasing in $(0, \infty)$. Then Ψ' is continuous in $[0, n^{-\rho}/c_n]$ and if

$$0 \leq x \leq y \leq \min\, \{x+\eta,\, n^\rho/c_n\},$$

then for some ξ between $c_n x$ and $c_n y$,

$$
\begin{aligned}
&|\Psi(c_n x) - \Psi(c_n y)| = |\Psi'(\xi)|\; |c_n x - c_n y| \\
&= O(\{Q'(\xi)/\xi\}^{1/2}\, c_n \eta) \quad \text{(by (V.14))} \\
&= O(\{Q'(n^\rho)/n^\rho\}^{1/2}\, c_n \eta) \quad \text{(by monotonicity of } Q'(t)/t) \\
&= O(n^{(\Omega-2)\rho/2\, +\, 1/\Lambda}\, \eta),
\end{aligned}
$$

by (II.2) and (II.12). Since Ψ is even, it follows that for $\eta > 0$,

$$\omega_n(\eta) \leq C_5\{n^{1/\Lambda-\rho}\, \eta + n^{-\Lambda\rho/2} + n^{(\Omega-2)\rho/2\, +\, 1/\Lambda}\, \eta\},$$

and in particular,

$$\omega_n(n^{-1/3}) \le C_5\{n^{1/\Lambda-\rho-1/3} + n^{-\Lambda\rho/2} + n^{(\Omega-2)\rho/2+1/\Lambda-1/3}\}.$$

We now choose $\rho := 2(1/3 - 1/\Lambda)/\{\Omega - 2 + \Lambda\}$, which ensures that $(\Omega-2)\rho/2 + 1/\Lambda - 1/3 = -\Lambda\rho/2$, and then

$$\Lambda\rho/2 > \Lambda\,(1/3 - 1/\Lambda)/\{2\Omega - 2\} =: \lambda. \tag{V.17}$$

Then

$$\omega_n(n^{-1/3}) = o(n^{-\lambda}),\ n \to \infty,$$

and by Jackson's Theorem, we can find polynomials R_n of degree at most $n^{1/3}$ such that uniformly for $x \in [-2,2]$,

$$|R_n(x) - \Psi(c_n x)| = o(n^{-\lambda}),\ n \to \infty.$$

Then as $\Psi \sim 1$ in $\mathbb{R}$, (V.10) follows. Further,

$$\|R_n\|_{L_\infty[-2,2]} = O(1),$$

so by Bernstein's inequality,

$$\|R_n'\|_{L_\infty[-1,1]} = O(n^{1/3}).$$

Also from (V.14), for $|x| \le 1$,

$$\left|\frac{d}{dx}\Psi(c_n x)\right| = O(c_n\,\{Q'(c_n x)/(c_n x)\}^{1/2})$$
$$= O(\{c_n Q'(c_n)\}^{1/2}) = O(n^{1/2}),$$

by (II.5) and (II.3). Using $\Psi \sim 1$, we then obtain (V.11). □

We can now prove an analogue of Theorem 13.2:

Theorem V.3

Let $W \in FR_\Lambda(3)$, some $\Lambda > 3$. Assume that $\{S_n\}_{n=1}^\infty$, $\{c_n\}_{n=1}^\infty$, $\{k_n\}_{n=1}^\infty$, $G_{Q/2}$, β, and L are as in Theorem V.1. Let

$$g_n(\theta) := |\sin\,\theta|^\beta\, W^2(c_n \cos\,\theta)/S_n(\cos\theta),\ \theta \in [-\pi,\pi]. \tag{V.18}$$

Then there exist n_1 and $\delta > 0$ such that uniformly for $n \ge n_1$, $\phi \in [n^{-\delta}, \pi - n^{-\delta}]$, and m satisfying (V.2), we have

$$|\phi_m(g_n,z)/\{z^m/\overline{D(g_n;z)}\} - 1| \le n^{-\delta},\ z:= e^{i\phi}. \tag{V.19}$$

Proof

This is **exactly** the same as that of Theorem 13.2, simply using Theorem V.1 and Lemma V.2 instead of Theorem 13.1 and Lemma 10.2 respectively. We leave it to the reader as an (easy) exercise. □

We must turn the asymptotics on the circle into asymptotics on the real line, but first must compare orthogonal polynomials for weights on $\mathbb{R}$ and their truncations: Following is an analogue of Lemma 13.3:

Lemma V.4

Let $W \in FR_\Lambda$, some $\Lambda > 0$. Let $K > 0$, and let $\{c_n\}_{n=1}^{\infty}$ be given by (I.12), with s large enough. Let

$$\Omega_n(x) := \begin{cases} W^2(c_nx), & x \in [-1,1], \\ 0 & , \text{otherwise}. \end{cases} \tag{V.20}$$

Then for $n \ge n_1$ and for $1 \le m \le n$,

$$\| c_n^{1/2} p_m(W^2,c_nx)W(c_nx) - p_m(\Omega_n,x)W(c_nx)\|_{L_\infty(\mathbb{R})} \le n^{-K}. \tag{V.21}$$

Proof

This is **exactly** the same as that of Lemma 13.3, the only difference being that we use (as our infinite–finite–range inequality) Theorem 7.2 in [14,p.50] (with there $K_n :=$ Ka, some large enough a), rather than Theorem 5.2 above. □

Theorem I.4 is the special case m = n of the following analogue of Theorem 13.4:

Theorem V.5

Let $W \in FR_\Lambda(3)$, some $\Lambda > 3$. Let $L > 0$, and let $\{c_n\}_{n=1}^{\infty}$ be given by (I.12) with s fixed but large enough. Let $\{\hat{F}_n\}_{n=1}^{\infty}$ be defined by (I.13). Then there exist n_1 and $\delta > 0$ s.t. for $n \geq n_1$,

$$c_n^{1/2}\, p_m(W^2,c_nx)\, \hat{F}_n(x)$$

$$= (\tfrac{2}{\pi})^{1/2}\, \mathrm{Re}\, [z^m\, \hat{F}_n(\cos\phi)/\overline{D(\hat{F}_n^2(\cos t);z)}] + O(n^{-\delta}), \tag{V.22}$$

uniformly for

$$x = \cos\phi,\ \phi \in [n^{-\delta},\pi - n^{-\delta}],\ z := e^{i\phi}, \tag{V.23}$$

and

$$n \geq m \geq n - \langle Ln\Delta_n \rangle. \tag{V.24}$$

We may also write

$$c_n^{1/2}\, p_m(W^2,c_nx)\, \hat{F}_n(x) = (\tfrac{2}{\pi})^{1/2}\cos(m\phi+\Gamma(\hat{F}_n^2;\phi)) + O(n^{-\delta}), \tag{V.25}$$

$$= (\tfrac{2}{\pi})^{1/2}\cos((m-n)\phi + n\pi \int_x^1 \mu_{n,c_n}(t)\, dt + \tfrac{\phi}{2} - \tfrac{\pi}{4}) + O(n^{-\delta}), \tag{V.26}$$

uniformly for x and m as in (V.23) and (V.24). Here μ_{n,c_n} is defined by (4.3) to (4.5).

<u>Proof</u>

This is the same as that of Theorem 13.4, but we provide the details. Let $\Omega_n(x)$ be defined by (V.20), and let

$$\mu_n'(\theta) := W^2(c_n\cos\theta)\,|\sin\theta| = \Omega_n(\cos\theta)\,|\sin\theta| = \hat{F}_n^2(\cos\theta),\ \theta \in [-\pi,\pi].$$

We have [30,p.294,eqn.(11.5.2)],

$$p_m(\Omega_n,x) = (2\pi)^{-1/2}\, \{1 + \phi_{2m}(\mu_n',0)/\kappa_{2m}(\mu_n')\}^{-1/2}$$

$$\times\ \{z^{-m}\, \phi_{2m}(\mu_n',z) + z^m\, \phi_{2m}(\mu_n',z^{-1})\}$$

$$= (2\pi)^{-1/2}\, \{1 + \phi_{2m}(\mu_n',0)/\kappa_{2m}(\mu_n')\}^{-1/2}$$

$$\times\ \{z^m/\overline{D(\mu_n';z)} + z^{-m}/\overline{D(\mu_n';1/z)}\}\ \{1 + O(n^{-\delta})\},$$

uniformly for x and m satisfying (V.23–24), by Theorem V.3, with $\beta = 1$ and $S_n = 1$.

Since

$$|D(\mu_n';1/z)| = |D(\mu_n';z)| = \mu_n'(\phi)^{1/2} = \hat{F}_n(x),$$

we have, again uniformly for x and m as above,

$$p_m(\Omega_n,x)\,\hat{F}_n(x) = (2\pi)^{-1/2}\,\{1+\phi_{2m}(\mu_n',0)/\kappa_{2m}(\mu_n')\}^{-1/2}$$

$$\times\,\{z^m\hat{F}_n(x)/\overline{D(\hat{F}_n^2(\cos\ t);z)} + z^{-m}\hat{F}_n(x)/\overline{D(\hat{F}_n^2(\cos\ t);\tfrac{1}{z})}$$

$$+\,O(n^{-\delta})\}.$$

It remains to show that

$$\Phi_{2m}(\mu_n',0) := \phi_{2m}(\mu_n',0)/\kappa_{2m}(\mu_n')$$

satisfies

$$\max\{\,|\Phi_{2m}(\mu_n',0)| : n - <Ln\Delta_n> \le m \le n\} = O(n^{-\delta}),\ n\to\infty, \tag{V.27}$$

and then to apply Lemma V.4 to complete the proof of (V.22). From (11.16) and (11.46) (with $R = c_n$), we can then deduce (V.25) and (V.26). There are several ways to establish (V.27), the simplest being to use an inequality of Rahmanov [28,p.150,Theorem 1]:

$$|\Phi_k(\mu_n',0)| \le 2\,[\inf_{P\in P_k} \frac{1}{2\pi}\int_{-\pi}^{\pi} \{|P(e^{i\theta})|\ \mu_n'(\theta)^{1/2} - 1\}^2\,d\theta]^{1/2}$$

$$\le 2\,[\inf_{\substack{S\in P_k\\ S\ge 0}} \frac{1}{2\pi}\int_{-\pi}^{\pi} \{[S(\cos\theta)W^2(c_n\cos\theta)\,|\sin\theta|\,]^{1/2} - 1\}^2 d\theta]^{1/2}. \tag{V.28}$$

Here the second inf is taken over all even trigonometric polynomials $S(\cos\theta)$ of degree at most k, non–negative in [–1,1]. The equivalence of $|P(e^{i\theta})|^2$ and $S(\cos\theta)$ is discussed in [30,p.3]. But according to Theorem IV.1, we can find polynomials P_n of degree at most $2n - 2Ln\Delta_n - n\Delta_n$, such that

$$P_n(t)/G_{Q/2}^2(c_nt)\,(1-t^2)^{1/2} \le 1,\ t\in(-1,1),$$

and for some $\delta > 0$,

$$|P_n(t)/G^2_{Q/2}(c_nt)\,(1-t^2)^{1/2}-1| \le n^{-2\delta},\ |t| \le 1-n^{-2\delta}.$$

Also, by Lemma V.2, we can find polynomials R_n of degree $\le n^{1/3} = o(n\Delta_n)$ such that for some $\lambda > 0$,

$$|R_n(t)W^2(c_nt)G^2_{Q/2}(c_nt)-1| \le C_1n^{-\lambda},\ |t| \le 1.$$

Then

$$S_n(t) := P_n(t)R_n(t)$$

is positive in $[-1,1]$, has degree at most $2n - Ln\Delta_n$, n large enough, and

$$0 < S_n(t)W^2(c_nt)\,(1-t^2)^{1/2} \le 1 + C_1\,n^{-\lambda},\ |t| \le 1,$$

while

$$|S_n(t)W^2(c_nt)\,(1-t^2)^{1/2}-1| \le C_3\,\{n^{-\lambda}+n^{-2\delta}\},$$

$|t| \le 1-n^{-2\delta}$. Then (V.28) and these last inequalities yield (V.27). □

In establishing asymptotics for Christoffel functions, we shall need an analogue of Theorem V.5 for the weight $1/G^2_{Q/2}$. Following is an analogue of Lemma 13.6:

<u>Lemma V.6</u>

Let $W \in FR_\Lambda$, some $\Lambda > 0$. Let H be an entire function as in Theorem III.1, with $\hat{J} = 0$. Let $K > 0$, and let $\{c_n\}_{n=1}^{\infty}$ be given by (I.12), with s large enough. Let

$$\Omega_n(x) := \begin{cases} 1/H^2(c_nx), & x \in [-1,1], \\ 0 & , \text{otherwise}. \end{cases} \tag{V.29}$$

Then for $n \ge n_1$ and for $1 \le m \le n$,

$$\|c_n^{1/2}p_m(H^{-2},c_nx)/H(c_nx) - p_m(\Omega_n,x)/H(c_nx)\|_{L_\infty(\mathbb{R})} \le n^{-K}. \tag{V.30}$$

<u>Proof</u>

Since $HW \sim 1$ in $\mathbb{R}$, (see (III.2)) we have

$$\|P/H\|_{L_p(|x|\ge c_n)} \le C_1\,\|PW\|_{L_p(|x|\ge c_n)}$$

$$\leq C_2\, n^{-2K}\, \|PW\|_{L_p(|x|\leq c_n)}$$

$$\leq C_3\, n^{-2K}\, \|P/H\|_{L_p(|x|\geq c_n)}, \tag{V.31}$$

for $n \geq n_1$, $p \in (0, \infty)$ fixed, and $P \in \mathcal{P}_n$, provided s is large enough. Using this inequality rather than Theorem 5.2 above, we can proceed as in Lemma 13.3. □

Following is our analogue of Theorem 13.7:

<u>Theorem V.7</u>

Let $W \in FR_\Lambda$, some $\Lambda > 0$. Let H be an entire function as in Theorem III.1, with $\hat{J} = 2$. Let $L > 0$, and let $\{c_n\}_{n=1}^{\infty}$ be given by (I.12), with s large enough. Let

$$\tilde{F}_n(x) := H^{-1}(c_n x)\,(1 - x^2)^{1/4},\ x \in (-1,1),\ n \geq 1. \tag{V.32}$$

Then there exist n_1 and $\delta > 0$ s.t. for $n \geq n_1$,

$$c_n^{1/2}\, p_m(H^{-2}, c_n x)\, \tilde{F}_n(x)$$

$$= (\tfrac{2}{\pi})^{1/2}\, \mathrm{Re}\, [z^m\, \tilde{F}_n(\cos\phi)/\overline{D(\tilde{F}_n^2(\cos t); z)}] + O(n^{-\delta}), \tag{V.33}$$

$$= (\tfrac{2}{\pi})^{1/2} \cos(m\phi + \Gamma(\tilde{F}_n^2; \phi)) + O(n^{-\delta}), \tag{V.34}$$

uniformly for x as in (V.23) and m as in (V.24).

<u>Proof</u>

This is exactly the same as that of Theorem V.5 (or Theorem 13.4 or Theorem 13.7). We merely substitute Theorem V.1 and Lemma V.6 for Theorem V.3 and Lemma V.4 respectively. □

VI. Christoffel Functions

In this section, we establish Theorems I.2 and I.3. The results and methods are very similar to those in Section 15, but we often provide the full details. First, an analogue of Theorem 15.1:

<u>Theorem VI.1</u>

Let $W \in FR_\Lambda$, some $\Lambda > 0$. Let H be an entire function as in Theorem III.1 with $\hat{J} = 2$. Let $\{c_n\}_{n=1}^{\infty}$ be defined by (I.12) with s fixed but large enough. Let

$$\tilde{F}_n(x) := H(c_n x)^{-1} (1 - x^2)^{1/4}, \; x \in (-1,1), \; n \geq 1. \tag{VI.1}$$

Then there exist n_1 and $\delta > 0$ such that uniformly for $n \geq n_1$ and

$$|x| \leq 1 - n^{-\delta}, \; x = \cos \phi, \tag{VI.2}$$

we have

$$\begin{aligned} &\pi \, \lambda_n^{-1}(H^{-2}, c_n x) \, H^{-2}(c_n x) \, (1 - x^2)^{1/2} \, c_n/n \\ &= 1 - \frac{1}{2n} + \{2n \sin \phi\}^{-1} \sin((2n-1)\phi + 2\Gamma(\tilde{F}_n^2;\phi)) + n^{-1}\Gamma'(\tilde{F}_n^2;\phi) \\ &+ O(n^{-1-\delta}). \end{aligned} \tag{VI.3}$$

<u>Proof</u>

We shall use Lemma 11.4, and the polynomials of Theorem IV.1. There will be three steps:

<u>Step 1</u> Lower Bounds

Let s in the definition of $\{c_n\}_{n=1}^{\infty}$ be chosen so large that if $\beta := \max\{1, 1/\Lambda\}$, then

$$\|P/H\|_{L_2(\mathbb{R})} \leq (1 + n^{-2\beta}) \, \|P/H\|_{L_2(|x| \leq c_n)}, \tag{VI.4}$$

$P \in \mathcal{P}_n$, $n \geq n_1$, this being possible by Theorem 7.2 in [14,p.50], and the fact that $H^{-1} \sim W$ in $\mathbb{R}$ (compare (V.31)). For $n \geq n_1$, let $P_n(u)$ be the polynomial of Theorem IV.1 (with $\kappa = 1/2$) of degree at most 2n–4, positive in [–1,1], except possibly for simple zeros at ± 1, and with the following properties: Let

$$\tau_n(u) := P_n(u)\, H^{-2}(c_n u)\, (1 - u^2)^{-1/2},\ u \in (-1,1).$$

Then there exists $\delta_1 > 0$ such that if $n \geq n_1$ and $|x_0| \leq 1 - n^{-\delta_1}$, then

$$\tau_n(x_0) = 1;\ \tau_n'(x_0) = 0, \tag{VI.5}$$

while

$$\tau_n(u) \leq 1,\ u \in [-1,1]. \tag{VI.6}$$

Furthermore (IV.6) to (IV.8) are valid. Then

$$H^{-2}(c_n u) \leq (1 - u^2)^{1/2}/P_n(u) =: \Omega_n(u),\ u \in (-1,1), \tag{VI.7}$$

and

$$H^{-2}(c_n x_0) = \Omega_n(x_0). \tag{VI.8}$$

Next, using (VI.4), we obtain (compare (15.9) above)

$$\lambda_n(H^{-2}, c_n x_0) \leq (1 + n^{-2\beta})^2 \inf_{P \in \mathcal{P}_{n-1}} \int_{-c_n}^{c_n} (P/H)^2(t)\, dt/P^2(c_n x_0)$$

$$\leq (1 + n^{-2\beta})^2\, c_n\, \lambda_n(\Omega_n, x_0). \tag{VI.9}$$

Next, if

$$g_n(x) := \Omega_n(x)\, (1 - x^2)^{1/2} = (1 - x^2)/P_n(x),\ x \in (-1,1), \tag{VI.10}$$

Lemma 11.4 shows that (15.10) holds and hence, using (VI.8) and (VI.9), we obtain for $|x_0| = |\cos \phi_0| \leq 1 - n^{-\delta_1}$,

$$\pi\, \lambda_n^{-1}(H^{-2}, c_n x_0)\, H^{-2}(c_n x_0)\, (1 - x_0^2)^{1/2}\, c_n$$

$$\geq \pi\, \lambda_n^{-1}(\Omega_n, x_0)\, \Omega_n(x_0)\, (1 - x_0^2)^{1/2}\, (1 + n^{-2\beta})^{-2}$$

$$= \{n - \tfrac{1}{2} + (2\sin \phi_0)^{-1} \sin((2n-1)\phi_0 + 2\Gamma(g_n;\phi_0))$$

$$+ \Gamma'(g_n;\phi_0)\}\, (1 + n^{-2\beta})^{-2}. \tag{VI.11}$$

Next, from (VI.1) and (VI.10),

$$\tilde{F}_n^2(x)/g_n(x) = H^{-2}(c_n x)\, P_n(x)\, (1 - x^2)^{-1/2} = \tau_n(x),$$

so that (compare (1.21)),

$$\Gamma(\tilde{F}_n^2;\phi) - \Gamma(g_n;\phi) = \Gamma(\tau_n;\phi). \tag{VI.12}$$

To estimate $\Gamma'(\tau_n;\phi_0)$ and $\Gamma(\tau_n;\phi_0)$, we use Lemma 11.3. Firstly, from (11.23),

$$|\Gamma(\tau_n;\phi_0)| \leq \frac{(1-x_0^2)^{1/2}}{2\pi} \int_{-1}^{1} \left| \frac{\log \tau_n(s) - \log \tau_n(x_0)}{s - x_0} \right| (1-s^2)^{-1/2}\, ds$$

$$= O(n^{-\delta_1} \log n),$$

from (IV.6) and (IV.7), by a straightforward estimation and provided δ_1 is small enough. The estimation of $\Gamma'(\tau_n;\phi_0)$ is harder, and uses Lemma 11.3(d). Choose there $\epsilon := n^{-\delta_1/2}$, to obtain

$$|\Gamma'(\tau_n;\phi_0)| \leq n^{\delta_1/2} \pi^{-1} \int_{-1}^{1} \left| \frac{\log \tau_n(s) - \log \tau_n(x_0)}{s - x_0} \right| (1-s^2)^{-1/2} ds$$

$$+ 2|\tau_n'(x_0)/\tau_n(x_0)| \log(2en^{\delta_1/2}) + (1-|x_0|)^{1/2} \int_0^{n^{-\delta_1/2}} \omega(\eta)/\eta\, d\eta.$$

Here, according to Lemma 11.3(d), for $\eta > 0$,

$$\omega(\eta) = \sup\{ |\tau_n'(s)/\tau_n(s) - \tau_n'(x_0)/\tau_n(x_0)| : |s - x_0| \leq \eta \}$$

$$= O(n^{-\delta_1}\eta),\ 0 \leq \eta \leq n^{-\delta_1},$$

by (IV.6) with $j = 2$, and by (VI.5). Hence, using (IV.6) and (VI.5), we obtain

$$|\Gamma'(\tau_n;\phi_0)| = O(n^{-\delta_1/2} \log n),$$

uniformly for $|x_0| = |\cos \phi_0| \leq 1 - 2n^{-\delta_1/2}$, and n large enough. If δ_1 is small enough, (VI.11) and (VI.12) then yield for $|x_0| \leq 1 - n^{-\delta_1}$,

$$\pi\, \lambda_n^{-1}(H^{-2}, c_n x_0)\, H^{-2}(c_n x_0)\, (1 - x_0^2)^{1/2}\, c_n$$

$$\geq (1 + n^{-2\beta})^{-2} \{n - \tfrac{1}{2} + (2\sin \phi_0)^{-1} \sin((2n-1)\phi_0 + 2\Gamma(\tilde{F}_n^2;\phi_0))$$

$$+ \Gamma'(\tilde{F}_n^2;\phi_0) + O(n^{-\delta_1})\}. \qquad \text{(VI.13)}$$

We remark here that the δ_1 was suitably small enough to take care of the term $(2\sin(\phi_0))^{-1}$.

<u>Step 2</u> $\Gamma'(\tilde{F}_n^2;\phi_0)$ is not so large.

To simplify (VI.13), we show that $\Gamma'(\tilde{F}_n^2;\phi_0)$ is not too large, using Lemma 11.3(d), and if necessary, making δ_1 still smaller. Now

$$|\log \tilde{F}_n^2(x)| \leq 2\,|\log H(c_n x)| + \tfrac{1}{2}\,|\log(1-x^2)|$$

$$\leq C_1 Q(c_n x) + \tfrac{1}{2}\,|\log(1-x^2)|$$

(by (III.2))

$$\leq C_2\, n + \tfrac{1}{2}\,|\log(1-x^2)|,$$

$|x| \leq 1$, by (II.4) and (II.5). Also, by (III.3), for $|x| \leq 1 - 1/n$,

$$\left|\tfrac{d}{dx} \log \tilde{F}_n^2(x)\right| \leq 2c_n\,|H'(c_n x)/H(c_n x)| + |x|/|1-x^2|$$

$$\leq C_3\{c_n Q'(\max\{1,c_n x\}) + (1-|x|)^{-1}\} \quad \text{(by (III.3))}$$

$$\leq C_3\{c_n Q'(c_n)\,(\max\{1,\, c_n x\}/c_n)^{\Lambda-1} + n\} \quad \text{(by (II.3))}$$

$$\leq C_4\, n \begin{Bmatrix} c_n^{1-\Lambda}, & \Lambda < 1 \\ 1, & \Lambda \geq 1 \end{Bmatrix} + C_4 n \leq C_5\, n^{\beta},$$

by (II.12). (Recall that $\beta = \max\{1,1/\Lambda\}$). Similarly, using (III.3), we obtain for some C_5 and C_6,

$$\left|\left(\tfrac{d}{dx}\right)^2 \log \tilde{F}_n(x)\right| \leq C_5\, n^{C_6},\ |x| \leq 1 - n^{-1}, \text{ say.}$$

Then from Lemma 11.3(d), we obtain, if $\delta_2 > 0$ is small enough,

$$|\Gamma'(\tilde{F}_n^2;\phi_0)| \leq C_7\, n^{\beta+\delta_2} \log n,\ |x_0| \leq 1 - n^{-\delta_2}. \qquad \text{(VI.14)}$$

Hence, we can rewrite (VI.13) as

$$\pi\, \lambda_n^{-1}(H^{-2}, c_n x_0)\, H^{-2}(c_n x_0)\, (1-x_0^2)^{1/2}\, c_n$$

$$\geq n - \frac{1}{2} + (2\sin \phi_0)^{-1}\sin((2n-1)\phi_0 + 2\Gamma(\tilde{F}_n^2;\phi_0))$$

$$+ \Gamma'(\tilde{F}_n^2;\phi_0) + O(n^{-\delta_1}), \tag{VI.15}$$

$$|x_0| = |\cos \phi_0| \leq 1 - n^{-\delta_2}.$$

Step 3 Upper Bounds

The proof of the upper bound corresponding to (VI.15) is somewhat easier, since we do not need to use an infinite–finite range inequality.

For $n \geq n_1$, let $\hat{P}_n(u)$ be the polynomial of Theorem IV.1 (with $\kappa = 1/2$) of degree at most 2n–4, positive in [–1,1], except possibly for simple zeros at ± 1, and with the following properties: Let

$$\tau_n(u) := \hat{P}_n(u)\, H^{-2}(c_n u)\, (1 - u^2)^{-1/2}, \quad u \in (-1,1).$$

Then (VI.5) holds for a given $|x_0| \leq 1 - n^{-\delta_1}$, (some fixed $\delta_1 > 0$), and

$$\tau_n(u) \geq 1, \quad u \in [-1,1],$$

while (IV.6) to (IV.8) hold. Then

$$H^{-2}(c_n u) \geq (1 - u^2)^{1/2}/\hat{P}_n(u) =: \hat{\Omega}_n(u), \quad u \in (-1,1),$$

and

$$H^{-2}(c_n x_0) = \hat{\Omega}_n(x_0).$$

Then (as in the proof of Theorem 15.1)

$$\lambda_n(H^{-2}, c_n x_0) \geq \inf_{P \in \mathcal{P}_{n-1}} \int_{-c_n}^{c_n} (P/H)^2(t)\, dt/P^2(c_n x_0)$$

$$\geq c_n\, \lambda_n(\hat{\Omega}_n, x_0).$$

Proceeding exactly as in Step 1, we obtain using Lemma 11.3 and Lemma 11.4,

$$\pi\, \lambda_n^{-1}(H^{-2}, c_n x_0)\, H^{-2}(c_n x_0)\, (1 - x_0^2)^{1/2}\, c_n$$

$$\leq \pi \lambda_n^{-1}(\hat{\Omega}_n, x_0)\, \hat{\Omega}_n(x_0)\, (1 - x_0^2)^{1/2}$$

$$= n - \frac{1}{2} + (2\sin \phi_0)^{-1} \sin((2n-1)\phi_0 + 2\Gamma(\tilde{F}_n^2; \phi_0))$$

$$+ \Gamma'(\tilde{F}_n^2; \phi_0) + O(n^{-\delta_1}),$$

for $|x_0| = |\cos \phi_0| \leq 1 - n^{-\delta_2}$. Together with (VI.15), this yields the result. □

We shall need an analogue of Lemma 15.2:

<u>Lemma VI.2</u>

Let $W \in FR_\Lambda(3)$, some $\Lambda > 0$, and let Q' be continuous at 0. Let $\{c_n\}_{n=1}^{\infty}$ be given by (I.12) with $s \geq 0$ fixed. Let $H := G_{Q/2}$, let Ψ be defined by (10.9), and for $n \geq 1$, let

$$Y_n := \max_{|u| \leq 1} \left|\frac{d}{du} \log \Psi(c_n u)\right|$$

$$= \max_{|u| \leq 1} |c_n H'(c_n u)/H(c_n u) - c_n Q'(c_n u)|. \qquad \text{(VI.16)}$$

Then

$$Y_n = O(n^{\max\{1/2, 1/\Lambda\}}),\ n \to \infty. \qquad \text{(VI.17)}$$

If in addition, $W \in SE(4)$, then

$$Y_n = O(n^{1/\Lambda}),\ n \to \infty. \qquad \text{(VI.18)}$$

<u>Proof</u>

Let $L > 0$ be fixed but large enough. From Theorem 6.1(b), if n is large enough (see (V.14)), and $L/c_n \leq x \leq 1$,

$$c_n H'(c_n x)/H(c_n x) - c_n Q'(c_n x) = O(c_n \{Q'(c_n x)/c_n x\}^{1/2})$$

$$= O(\{c_n Q'(c_n)\, x^{\Lambda-2}\}^{1/2}) \qquad \text{(by (II.3))}$$

$$= O(n^{1/2}\, x^{(\Lambda-2)/2}),$$

by (II.5). Hence

$$\max_{L/c_n \le |u| \le 1} |c_n H'(c_n x)/H(c_n x) - c_n Q'(c_n x)|$$

$$= \begin{cases} O(n^{1/2}) & , \text{ if } \Lambda \ge 2, \\ O(n^{1/2} c_n^{-(\Lambda-2)/2}), & \text{if } \Lambda < 2, \end{cases}$$

$$= O(n^{\max\{1/2,1/\Lambda\}}),$$

by (II.12). Also as Q' and H'/H are continuous at 0,

$$\max_{|u| \le L/c_n} |c_n H'(c_n x)/H(c_n x) - c_n Q'(c_n x)| = O(c_n) = O(n^{1/\Lambda}).$$

Hence (VI.17). If in addition $W \in FR_\Lambda(4)$, then Theorem 6.1(c) yields

$$H'(t)/H(t) - Q'(t) = O(Q'(t)/Q(t) = O(1/t), \ t \to \infty,$$

by (II.4). Hence if $L/c_n \le x \le 1$,

$$|c_n H'(c_n x)/H(c_n x) - c_n Q'(c_n x)| = O(1/x) = O(c_n).$$

Then (VI.18) follows. □

The next step in obtaining asymptotics for the Christoffel functions $\lambda_n(W^2,x)$ is to respectively replace c_n by a_n and $\tilde{F}_n$ by F_n in (VI.3). Following is our analogue of Theorem 15.3.

<u>**Theorem VI.3**</u>

Let $W \in FR_\Lambda(3)$, some $\Lambda > 1$. Let $H := G_{Q/2}$ be defined by (1.15), let μ_{n,a_n} be defined by (2.20), and let

$$F_n(x) := W(a_n x)\ (1 - x^2)^{1/4}, \ x \in (-1,1), \ n \ge 1. \tag{VI.19}$$

Let $\eta > 0$. Then there exists $\delta_1 > 0$ such that for

$$|x| \le 1 - n^{-\delta_1}, \ x = \cos\phi, \ z = e^{i\phi}, \tag{VI.20}$$

we have

$$\pi\, \lambda_n^{-1}(H^{-2},a_n x)\ H^{-2}(a_n x)\ (1 - x^2)^{1/2}\ a_n/n$$

$$= 1 + n^{-1}\Gamma'(F_n^2;\phi) + O(\epsilon_n), \tag{VI.21}$$

$$= 1 + n^{-1}\ \mathrm{Re}[zD'(F_n^2(\cos t);z)/D(F_n^2(\cos t);z)] + O(\epsilon_n), \tag{VI.22}$$

$$= \pi\ \mu_{n,a_n}(x)\ (1 - x^2)^{1/2} + O(\epsilon_n), \tag{VI.23}$$

where

$$\epsilon_n := n^{\eta + \max\{-1/2,1/\Lambda-1\}},\ n \geq 1. \tag{VI.24}$$

If in addition, $W \in FR_\Lambda(4)$, then we may choose

$$\epsilon_n := n^{\eta + \max\{-2/3,1/\Lambda-1\}},\ n \geq 1. \tag{VI.25}$$

<u>Proof</u>

This will consist of two steps, each transforming (VI.3).

<u>Step 1</u> From $\tilde{F}_n$ to $\hat{F}_n$.

We replace $\tilde{F}_n$ in (VI.3) by $\hat{F}_n$, where

$$\hat{F}_n(x) := W(c_n x)\ (1 - x^2)^{1/4},\ x \in (-1,1), \tag{VI.26}$$

To this end, we use Lemma 11.3(d) to estimate

$$\Gamma'(\tilde{F}_n^2;\phi) - \Gamma'(\hat{F}_n^2;\phi) = \Gamma'(\tilde{F}_n^2/\hat{F}_n^2;\phi). \tag{VI.27}$$

Note that from (VI.1) and (VI.26),

$$(\tilde{F}_n^2/\hat{F}_n^2)(x) = \{WG_{Q/2}\}^{-2}(c_n x) = \Psi^2(c_n x),$$

where Ψ is given by (10.9). Now uniformly for $|x| \leq 1$, Theorem 6.1(a) and the fact that $T \sim 1$ in $\mathbb{R}$ show that

$$|\log \Psi(c_n x)| = O(1), \tag{VI.28}$$

while uniformly for $|x| \leq 1$, Lemma VI.2 shows that

$$\frac{d}{dx} \log \Psi(c_n x) = O(Y_n). \tag{VI.29}$$

We note that since $\Lambda > 1$, $Q'(x) \to 0$ as $x \to 0$ (by (II.1)) and so Q' is continuous at 0. Finally, it is not difficult to see from Theorem 6.1(a) that given $L > 0$, then for some C_3 and C_4,

$$\left|\left(\frac{d}{dx}\right)^2 \log \Psi(c_n x)\right| \le C_3 n^{C_4}, \quad L/c_n \le |x| \le 1. \tag{VI.30}$$

Next, Q' satisfies a Lipschitz condition of order $\Lambda - 1$ near 0, for if s, t $\in [0,1]$,

$$|Q'(s) - Q'(t)| = \left|\int_s^t Q''(u)du\right|$$

$$\le C_1 \left|\int_s^t Q'(u)/u \, du\right| \le C_2 \left|\int_s^t u^{\Lambda-2} \, du\right|$$

(by (II.9) and (II.1))

$$= C_3 |t^{\Lambda-1} - s^{\Lambda-1}| \le C_4 |s - t|^{\Lambda-1}.$$

Since $G_{Q/2}$ is positive and entire, we see then that for some C_5, C_6,

$$\left|\frac{d}{dx}\log \Psi(c_n x) - \frac{d}{dy}\log \Psi(c_n y)\right| \le C_5 n^{C_6} |x-y|^{\Lambda-1}, \ |x|, |y| \le L/c_n. \tag{VI.31}$$

Next, by Lemma 11.3(d), for given $\delta_2 > 0$,

$$|\Gamma'(\tilde{F}_n^2/\hat{F}_n^2;\phi)| \le 2n^{\delta_2}/\pi \int_{-1}^{1} \left|\frac{\log \Psi(c_n s) - \log \Psi(c_n x)}{s-x}\right| (1-s^2)^{-1/2} \, ds$$

$$+ 2\left|\frac{d}{dx}\log \Psi(c_n x)\right| \log(2en^{\delta_2}) + (1-|x|)^{1/2} \int_0^{n^{-\delta_2}} \omega(\eta)/\eta \, d\eta,$$

where for $\eta > 0$,

$$\omega(\eta) := \sup\left\{ \left|\frac{d}{dx}\log \Psi^2(c_n x) - \frac{d}{dt}\log \Psi^2(c_n t)\right| : |x-t| \le \eta\right\}.$$

Combining (VI.28) to (VI.31), we obtain

$$|\Gamma'(\tilde{F}_n^2/\hat{F}_n^2;\phi)| \le C_7 \{n^{\delta_2} \log n + Y_n \log n\}$$

$$\le C_8 \log n \, \{n^{\delta_2} + Y_n\},$$

provided

$$|x| = |\cos \phi| \le 1 - n^{-\delta_1}, \text{ some } \delta_1 > 0. \tag{VI.32}$$

Note that δ_2 may be made as small as we please, while δ_1 depends on δ_2. Of course, we may assume that δ_1 is as small as we please (but fixed), since then the range

(VI.32) of x becomes smaller. Then from (VI.3), we obtain uniformly for the range (VI.32),

$$\pi\,\lambda_n^{-1}(H^{-2},c_n x)\,H^{-2}(c_n x)\,(1-x^2)^{1/2}\,c_n/n$$

$$= 1 + n^{-1}\,\Gamma'(\hat{F}_n^2;\phi) + O(\frac{\log\, n}{n}\,\{n^{\delta_2} + Y_n\}). \qquad \text{(VI.33)}$$

<u>**Step 2**</u> From c_n to a_n and $\hat{F}_n$ to F_n

We now make the substitution

$$c_n x = a_n u \qquad \text{(VI.34)}$$

in (VI.33), using Lemma 11.5(b). Let us write

$$x = \cos\,\phi;\quad u = \cos\,\psi. \qquad \text{(VI.35)}$$

Note that in the notation (11.44), $\hat{F}_n$ and F_n become

$$\hat{F}_n = F_{c_n};\quad F_n = F_{a_n}.$$

Thus we may apply Lemma 11.5(b) with $R_1 = a_n$; $R_2 = c_n$; $x_1 = u$; $x_2 = x$; $\phi_1 = \psi$; $\phi_2 = \phi$, and

$$\beta = a_n/c_n = (1 + s\Delta_n)^{-1}, \qquad \text{(VI.36)}$$

to deduce that

$$|\Gamma'(F_n^2;\psi) - \Gamma'(\hat{F}_n^2;\phi)|$$

$$\le \frac{4}{\pi}(1-\beta)\int_0^\beta \left|\frac{c_n sQ'(c_n s) - c_n xQ'(c_n x)}{s^2 - x^2}\right|(1-s^2)^{-1/2}\,ds$$

$$+ \frac{2}{\pi}\int_\beta^1 \left|\frac{c_n sQ'(c_n s) - c_n xQ'(c_n x)}{s^2 - x^2}\right|(1-s^2)^{1/2}\,ds$$

$$\le C_9\,\{\Delta_n \int_0^\beta \left|\frac{c_n sQ'(c_n s) - c_n xQ'(c_n x)}{c_n s - c_n|x|}\right|\frac{c_n}{s+|x|}(1-s^2)^{-1/2}\,ds$$

$$+ \max_{s\in[\beta,1]}\left|\frac{c_n sQ'(c_n s) - c_n xQ'(c_n x)}{c_n s - c_n|x|}\right| c_n\,\Delta_n^{3/2}\}. \qquad \text{(VI.37)}$$

Now for some ξ between $|x|$ and s,

$$\left|\frac{c_n sQ'(c_n s) - c_n xQ'(c_n x)}{c_n s - c_n |x|}\right| = |\frac{d}{du}(uQ'(u))|_{u=c_n\xi}|$$

$$= T(c_n\xi)\ |Q'(c_n\xi)| \le C_{10}\ Q'(c_n) \le C_{11} n/a_n,$$

since $T \sim 1$ and by monotonicity of $Q'(t)$ in $(0,\infty)$ – see (II.8) and (II.9). Then if also $x \neq 0$,

$$|\Gamma'(F_n^2;\psi) - \Gamma'(\hat{F}_n^2;\phi)| \le C_{12}\{\Delta_n n \int_0^\beta \frac{1}{s+|x|}(1-s^2)^{-1/2}\,ds + n\Delta_n^{3/2}\}$$

$$\le C_{13}\ \{(\log n)^{2/3}\ n^{1/3}\ \log[2 + \frac{1}{|x|}] + \log n\}$$

(by definition (I.11) of Δ_n)

$$\le C_{14}\ n^{1/3}\ (\log n)^{5/3}, \tag{VI.38}$$

provided $|x| \ge n^{-L}$, any fixed $L > 0$. If $|x| \le n^{-L}$, then we need estimate only the integral in the right–hand side of (VI.37) a little differently. As above, the integral over the range $[n^{-L},\beta]$ admits the estimate (VI.38). We must estimate the integral over $[0,n^{-L}]$. Now as $c_n^\Lambda n^{-L} = o(1)$ if L is large enough, and as $tQ'(t)$ is even, and increasing in $(0,\infty)$,

$$\Delta_n \int_{2|x|}^{2n^{-L}} \left|\frac{c_n sQ'(c_n s) - c_n xQ'(c_n x)}{c_n s - c_n |x|}\right| \frac{c_n}{s+|x|}(1-s^2)^{-1/2}\,ds$$

$$\le \Delta_n \int_{2|x|}^{2n^{-L}} \frac{c_n sQ'(c_n s)}{c_n s/2}\,\frac{c_n}{s+|x|}(1-s^2)^{-1/2}\,ds$$

$$\le \Delta_n \int_{2|x|}^{2n^{-L}} C_{16}\ (c_n s)^{\Lambda-2}\ c_n^2\ (1-s^2)^{-1/2}\,ds = o(\Delta_n),$$

by (II.1) and if L is large enough. Also, as above,

$$\Delta_n \int_0^{2|x|} \left|\frac{c_n sQ'(c_n s) - c_n xQ'(c_n x)}{c_n s - c_n |x|}\right| \frac{c_n}{s+|x|}(1-s^2)^{-1/2}\,ds$$

$$\leq C_{17}\, \Delta_n\, Q'(c_n)\, c_n^{\,2|x|/|x|}$$

$$\leq C_{18}\, n^{1/3}\, (\log n)^{2/3},$$

as above. Hence (VI.38) persists for $|x| \leq n^{-L}$. Combining (VI.33) and (VI.38) leads to (recall we have $c_n x = a_n u$)

$$\pi\, \lambda_n^{-1}(H^{-2}, a_n u)\, H^{-2}(a_n u)\, (1 - (a_n u/c_n)^2)^{1/2}\, c_n/n$$

$$= 1 + n^{-1}\, \Gamma'(F_n^2;\psi) + O(n^{-2/3}(\log n)^{5/3})$$

$$+ O(\frac{\log n}{n}\, \{n^{\delta_2} + Y_n\}), \tag{VI.39}$$

for the range (VI.32) of $x = a_n u/c_n$. Now since $a_n/c_n = 1 + O(\Delta_n)$,

$$(1 - (a_n u/c_n)^2)^{1/2}/(1 - u^2)^{1/2} = 1 + O(\Delta_n(1 - u^2)^{-1})$$

$$= 1 + O(\Delta_n\, n^{\delta_1}),$$

for the range (VI.32) of $x = a_n u/c_n$. Also, exactly as in step 2 of the proof of Theorem VI.1 (see (VI.14)), given a small enough $\delta_2 > 0$, we obtain from Lemma 11.3(d) (recall $\Lambda > 1$),

$$|\Gamma'(F_n^2;\psi)| \leq C_{22}\, n^{1+\delta_2},$$

for the range (VI.32) of $x = a_n u/c_n$, provided δ_1 is small enough. It follows that if δ_1, δ_2 are small enough, we may rewrite (VI.39) as

$$\pi\, \lambda_n^{-1}(H^{-2}, a_n u)\, H^{-2}(a_n u)\, (1 - u^2)^{1/2}\, a_n/n$$

$$= 1 + n^{-1}\Gamma'(F_n^2;\psi) + O(n^{-2/3+\delta_1+\delta_2}(\log n)^{5/3}) + O(\frac{\log n}{n} Y_n), \tag{VI.40}$$

uniformly for

$$|u| \leq 1 - n^{-\delta_1},\ u = \cos\psi, \tag{VI.41}$$

where δ_1 may be made arbitrarily small. Now by the estimates for Y_n in Lemma VI.2, if δ_1, δ_2 are small enough, and $\eta > 0$ is given,

$$O(n^{-2/3+\delta_1+\delta_2}(\log n)^{5/3}) + O(\frac{\log n}{n} Y_n)$$

$$= \begin{cases} O(n^{\eta+\max\{-1/2,1/\Lambda-1\}}), & W \in FR_\Lambda(3), \\ O(n^{\eta+\max\{-2/3,1/\Lambda-1\}}), & W \in FR_\Lambda(4). \end{cases}$$

Finally, by (11.45),

$$1 + n^{-1}\Gamma'(F_n^2;\psi)$$

$$= \pi \sin \psi \, \mu_{n,a_n}(\cos \psi) + \frac{1}{2n}$$

$$= 1 + n^{-1}\mathrm{Re}[e^{i\psi}D'(F_n^2(\cos t);e^{i\psi})/D(F_n^2(\cos t);e^{i\psi})],$$

– compare (11.18). On changing u to x, and ψ to ϕ, we have the stated result. □

To turn the asymptotics for $\lambda_n(H^{-2},a_n x)$ into those for $\lambda_n(W^2,x)$, we shall use (as in Section 15) Nevai's operators [21,6,8]. Recall first if $H = 1/G_{Q/2}$, the notation (15.41) to (15.42), and Nevai's inequality (15.43). We need the following analogue of Lemma 15.4:

Lemma VI.4

Let $W \in FR_\Lambda$, some $\Lambda > 1$, and let $H = G_{Q/2}$ be defined by (1.15).

(a) There exists $\delta > 0$ such that

$$(1-x^2)^{1/4} (1-t^2)^{1/4} \, |K_n(H^{-2},a_n x,a_n t)| \; H^{-1}(a_n x)H^{-1}(a_n t)$$

$$\leq C_1/\{a_n |x-t|\}, \quad |x|, |t| \leq 1 - n^{-\delta}. \tag{VI.42}$$

Further, for $t \in \mathbb{R}$, and $|x| \leq 1 - n^{-\delta}$,

$$(1-x^2)^{1/4} \, |K_n(H^{-2},a_n x,a_n t)| \; H^{-1}(a_n x)$$

$$\leq C_2 \{ \sum_{j=n-1}^{n} |p_j(H^{-2},a_n t)| \}/\{ a_n^{1/2} \, |x-t| \}. \tag{VI.43}$$

(b) There exists $\delta > 0$ such that

$$\lambda_n(H^{-2},a_n x) \, H^2(a_n x) \, (1 - x^2)^{1/2} \sim a_n/n, \tag{VI.44}$$

for $|x| \leq 1 - n^{-\delta}$.

(c) There exists $\delta > 0$ such that

$$|K_n(H^{-2},a_nx,a_nt)|\ H^{-1}(a_nx)H^{-1}(a_nt)\ (1-x^2)^{-1/4}(1-t^2)^{-1/4}$$
$$\leq C_5\ n/a_n, \tag{VI.45}$$

$|x|, |t| \leq 1 - n^{-\delta}$.

Proof

(a) Exactly as in Lemma 15.3, (15.48) holds. (We can use (V.31) for the infinite–finite range inequality, instead of Lemma 13.5(a)). Next, from Theorem V.7,

$$|c_n^{1/2}\ p_j(H^{-2},c_nt)\ H^{-1}(c_nt)\ (1-t^2)^{1/4}| \leq 2, \tag{VI.46}$$

for $j = n-1$, n, and $|t| \leq 1 - n^{-\delta_2}$, some $\delta_2 > 0$. Then (15.41), (15.48) and (VI.46) imply (VI.42) and (VI.43).

(b) From Theorem VI.3, for some small fixed $\eta > 0$,

$$\lambda_n^{-1}(H^{-2},a_nx)\ H^{-2}(a_nx)\ a_n/n$$
$$= \mu_{n,a_n}(x) + O(n^{\eta+\max\{-1/2,1/\Lambda-1\}}(1-|x|)^{-1/2}), \tag{VI.47}$$

uniformly for $|x| \leq 1 - n^{-\delta}$, some $\delta > 0$. We note that combining (5.49) in [14,p.38], (6.24) and (6.26) in [14,p.44], we see that there exists $\epsilon_0 > 0$ such that

$$\mu_{n,a_n}(x) = \nu_{n,a_n}(x) \sim (1-|x|)^{1/2},\ |x| \in [1-\epsilon_0,1].$$

It is also not difficult to see from (2.20), (II.3), (II.5) and the fact that $\Lambda > 1$, that

$\mu_{n,a_n}(x) \sim 1$, $|x| \leq 1 - \epsilon_0$. (Compare the proof of Lemma 15.4 or Lemma 3.1 in [9]). Then (VI.47) yields (VI.44).

(c) By the Cauchy–Schwarz inequality,

$$K_n(H^{-2},a_nx,a_nt) \leq \{\sum_{j=0}^{n-1} p_j^2(H^{-2},a_nx)\}^{\frac{1}{2}} \{\sum_{j=0}^{n-1} p_j^2(H^{-2},a_nt)\}^{\frac{1}{2}}$$
$$= \lambda_n^{-1/2}(H^{-2},a_nx)\ \lambda_n^{-1/2}(H^{-2},a_nt).$$

Then (VI.44) yields (VI.45). □

Finally we can prove Theorem I.2:

Theorem VI.5

Let $W \in FR_\Lambda(3)$ and let $\eta > 0$. Then there exists $\delta > 0$ such that uniformly for

$$|x| \le 1 - n^{-\delta};\ x = \cos\phi;\ z = e^{i\phi}, \tag{VI.48}$$

we have

$$\pi\,\lambda_n^{-1}(W^2,a_nx)\,W^2(a_nx)\,(1-x^2)^{1/2}\,a_n/n$$

$$= 1 + n^{-1}\,\mathrm{Re}[zD'(F_n^2(\cos t);z)/D(F_n^2(\cos t);z)] + O(\epsilon_n), \tag{VI.49}$$

$$= 1 + n^{-1}\Gamma'(F_n^2;\phi) + O(\epsilon_n), \tag{VI.50}$$

$$= \pi\,\mu_{n,a_n}(x)\,(1-x^2)^{1/2} + O(\epsilon_n), \tag{VI.51}$$

where

$$\epsilon_n := n^{\eta + \max\{-1/2,-1+1/\Lambda\}},\ n \ge 1. \tag{VI.52}$$

If in addition $W \in SE(4)$, then we may take

$$\epsilon_n := n^{\eta + \max\{-2/3,-1+1/\Lambda\}},\ n \ge 1. \tag{VI.53}$$

Proof

We let $H := G_{Q/2}$ for notational simplicity. Recall from (10.9), the definition of Ψ. Then Nevai's inequality (15.43) implies that (compare (15.57))

$$\{\mathscr{G}_n[\Psi^2](a_nx)/\Psi^2(a_nx)\}^{-1} \le \frac{\lambda_n(W^2,a_nx)\,W^{-2}(a_nx)}{\lambda_n(G_{Q/2}^{-2},a_nx)\,G_{Q/2}^2(a_nx)}$$

$$\le \mathscr{G}_n[\Psi^{-2}](a_nx)/\Psi^{-2}(a_nx). \tag{VI.54}$$

Let $\sigma = \pm 2$, and let $\delta > 0$. Further, let $\{\theta_n\}_{n=1}^{\infty}$ be a strictly decreasing sequence of positive numbers with limit 0. Since $\mathscr{G}_n[1] = 1$, (see [8,21]), we have

$$\mathscr{G}_n[\Psi^\sigma](a_nx)/\Psi^\sigma(a_nx) - 1$$

$$= \lambda_n(H^{-2},a_nx) \int_{-\infty}^{\infty} K_n^2(H^{-2},a_nx,a_nt) \left[\frac{\Psi^\sigma(a_n t)-\Psi^\sigma(a_nx)}{\Psi^\sigma(a_nx)}\right] a_nH^{-2}(a_nt)\, dt$$

$$= \lambda_n(H^{-2},a_nx) \left\{\int_{|x-t|\le\theta_n} + \int_{\substack{|x-t|\ge\theta_n \\ |t|\le 1-n^{-\delta}}} + \int_{|t|\ge 1-n^{-\delta}}\right\}$$

$$\times\ K_n^2(H^{-2},a_nx,a_nt) \left[\frac{\Psi^\sigma(a_n t)-\Psi^\sigma(a_nx)}{\Psi^\sigma(a_nx)}\right] a_nH^{-2}(a_nt)\, dt$$

$$=: I_1 + I_2 + I_3. \qquad \text{(VI.55)}$$

Now by (VI.44) and (VI.45), and the fact that $\Psi \sim 1$ in $\mathbb{R}$,

$$I_1 \le C_6\, n \int_{|x-t|\le\theta_n} \left|\Psi^\sigma(a_nt) - \Psi^\sigma(a_nx)\right|\, dt. \qquad \text{(VI.56)}$$

Now for $|x| \le 1-\theta_n-n^{-\delta}$, and t such that $|x-t| \le \theta_n$,

$$|\Psi^\sigma(a_nt) - \Psi^\sigma(a_nx)| \le |t-x| \max_{|u|\le 1} \left|\frac{d}{du}\Psi^\sigma(a_nu)\right|$$

$$= O(\theta_nY_n),$$

with the notation of Lemma VI.2 and since $\Psi \sim 1$ in $\mathbb{R}$. Then (VI.56) yields

$$I_1 \le C_7\, n\, \theta_n^2 Y_n. \qquad \text{(VI.57)}$$

Next, using (VI.42) and (VI.44),

$$I_2 \le C_8\, n^{-1}\, (1-x^2)^{-1}$$

$$\times \int_{\substack{|x-t|\ge\theta_n \\ |t|\le 1-n^{-\delta}}} |x-t|^{-2}\, |\Psi^\sigma(a_nt) - \Psi^\sigma(a_nx)|\, (1-t^2)^{-1/2}\, dt.$$

Here as above,

$$|\Psi^\sigma(a_nt) - \Psi^\sigma(a_nx)| = O(|t-x|\, Y_n),$$

and hence for $|x| \le 1 - n^{-\delta}$, some small enough δ,

$$I_2 \le C_8\, n^{-1+3\delta/2}\, Y_n \log(1/\theta_n). \qquad \text{(VI.58)}$$

Finally, using (VI.43) and (VI.44), we obtain for $|x| \le 1 - 2n^{-\delta}$,

$$I_3 \le C_9\, n^{-1+\delta} \sum_{j=n-1}^{n} \int_{|t|\ge 1-n^{-\delta}} p_j^2(H^{-2}, a_n t) H^{-2}(a_n t)$$

$$\times\, |x-t|^{-2}\, a_n\, dt.$$

$$\le C_{10}\, n^{-1+3\delta}, \tag{VI.59}$$

since $|x-t| \ge 1 - n^{-\delta} - |x| \ge n^{-\delta}$. Hence if $|x| \le 1 - 2n^{-\delta}$, for small enough $\delta >$ 0, we obtain from (VI.55), (VI.57), (VI.58), (VI.59),

$$|\mathcal{G}_n[\Psi^\sigma](a_n x)/\Psi^\sigma(a_n x) - 1|$$

$$\le C_{11}\{n\, \theta_n^2\, Y_n + n^{-1+3\delta/2}\, Y_n \log(1/\theta_n) + n^{-1+3\delta}\}.$$

Note from Lemma VI.2 that we may assume that Y_n grows at least as fast as some positive power of n (else replace it by the relevant upper bound of Lemma VI.2). Then choosing

$$\theta_n := \min\{\, n^{-100},\, Y_n^{-1}\, (\log n)^{-1}\},$$

we obtain from (VI.17),

$$|\mathcal{G}_n[\Psi^\sigma](a_n x)/\Psi^\sigma(a_n x) - 1|$$

$$\le C_{13}\{n^{-1+3\delta/2}\, Y_n \log n + n^{-1+3\delta}\}.$$

$$\le C_{14}\, n^{2\delta + \max\{-1/2,-1+1/\Lambda\}},$$

if δ is small enough and $|x| \le 1 - 2n^{-\delta}$. Then from Theorem VI.3, and (VI.54),

$$\pi\, \lambda_n^{-1}(W^2, a_n x)\, W^2(a_n x)\, (1-x^2)^{1/2}\, a_n/n$$

$$= \{\pi\, \mu_{n,a_n}(x)(1-x^2)^{1/2} + O(n^{\delta+\max\{-1/2,1/\Lambda-1\}})\}$$

$$\times\, \{1 + O(n^{2\delta+\max\{-1/2,1/\Lambda-1\}})\}$$

$$= \pi\, \mu_{n,a_n}(x)\, (1-x^2)^{1/2} + O(n^{2\delta+\max\{-1/2,1/\Lambda-1\}}),$$

since $\mu_{n,a_n}(x)$ is bounded (see the proof of Lemma VI.4). Thus we have the result if

$W \in FR_\Lambda(3)$. If in addition $W \in FR_\Lambda(4)$, then we use instead the better estimate (VI.18) for Y_n and the better order estimate in Theorem VI.3 to obtain the result. □

Index